Microprocessor 5

Series Editor
Jean-Charles Pomerol

Microprocessor 5

Software and Hardware Aspects of Development, Debugging and Testing – The Microcomputer

Philippe Darche

iSTE

WILEY

First published 2020 in Great Britain and the United States by ISTE Ltd and John Wiley & Sons, Inc.

ISTE Ltd
27-37 St George's Road
London SW19 4EU
UK

www.iste.co.uk

John Wiley & Sons, Inc.
111 River Street
Hoboken, NJ 07030
USA

www.wiley.com

Library of Congress Control Number: 2020943158

British Library Cataloguing-in-Publication Data
A CIP record for this book is available from the British Library
ISBN 978-1-78630-651-7

Contents

Quotation

Every advantage has its disadvantages and vice versa.

Shadokian philosophy[1]

1 The Shadoks are the main characters from an experimental cartoon produced by the Research Office of the Office de Radiodiffusion-Télévision Française (ORTF). The two-minute-long episodes of this daily cult series were broadcast on ORTF's first channel (the only one at the time!) beginning in 1968. The birds were drawn simply and quickly using an experimental device called an *animograph*.

The Shadoks are ridiculous, stupid and mean. Their intellectual capacities are completely unusual. For example, they are known for bouncing up and down, but it is not clear why! Their vocabulary consists of four words: GA, BU, ZO and MEU, which are also the four digits in their number system (base 4) and the musical notes in their four-tone scale. Their philosophy is comprised of famous mottos such as the one cited in this book.

Preface

Computer systems (hardware and software) are becoming increasingly complex, embedded and transparent. It therefore is becoming difficult to delve into basic concepts in order to fully understand how they work. In order to accomplish this, one approach is to take an interest in the history of the domain. A second way is to soak up technology by reading datasheets for electronic components and patents. Last but not least is reading research articles. I have tried to follow all three paths throughout the writing of this series of books, with the aim of explaining the hardware and software operations of the microprocessor, the modern and integrated form of the central unit.

About the book

This five-volume series deals with the general operating principles of the microprocessor. It focuses in particular on the first two generations of this programmable component, that is, those that handle integers in 4- and 8-bit formats. In adopting a historical angle of study, this deliberate decision allows us to return to its basic operation without the conceptual overload of current models. The more advanced concepts, such as the mechanisms of virtual memories and cache memory or the different forms of parallelism, will be detailed in a future book with the presentation of subsequent generations, that is, 16-, 32- and 64-bit systems.

The first volume addresses the field's introductory concepts. As in music theory, we cannot understand the advent of the microprocessor without talking about the history of computers and technologies, which is presented in the first chapter. The second chapter deals with storage, the second function of the computer present in the microprocessor. The concepts of computational models and computer architecture will be the subject of the final chapter.

The second volume is devoted to aspects of communication in digital systems from the point of view of buses. Their main characteristics are presented, as well as their communication, access arbitration, and transaction protocols, their interfaces and their electrical characteristics. A classification is proposed and the main buses are described.

The third volume deals with the hardware aspects of the microprocessor. It first details the component's external interface and then its internal organization. It then presents the various commercial generations and certain specific families such as the Digital Signal Processor (DSP) and the microcontroller. The volume ends with a presentation of the datasheet.

The fourth volume deals with the software aspects of this component. The main characteristics of the Instruction Set Architecture (ISA) of a generic component are detailed. We then study the two ways to alter the execution flow with both classic and interrupt function call mechanisms.

The final volume presents the hardware and software aspects of the development chain for a digital system as well as the architectures of the first microcomputers in the historical perspective.

Multi-level organization

This book gradually transitions from conceptual to physical implementation. Pedagogy was my main concern, without neglecting formal aspects. Reading can take place on several levels. Each reader will be presented with introductory information before being asked to understand more difficult topics. Knowledge, with a few exceptions, has been presented linearly and as comprehensively as possible. Concrete examples drawn from former and current technologies illustrate the theoretical concepts.

When necessary, exercises complete the learning process by examining certain mechanisms in more depth. Each volume ends with bibliographic references including research articles, works and patents at the origin of the concepts and more recent ones reflecting the state of the art. These references allow the reader to find additional and more theoretical information. There is also a list of acronyms used and an index covering the entire work.

This series of books on computer architecture is the fruit of over 30 years of travels in the electronic, microelectronic and computer worlds. I hope that it will provide you with sufficient knowledge, both practical and theoretical, to then specialize in one of these fields. I wish you a pleasant stroll through these different worlds.

IMPORTANT NOTES.– As this book presents an introduction to the field of microprocessors, references to components from all periods are cited, as well as references to computers from generations before this component appeared.

Original company names have been used, although some have merged. This will allow readers to find specification sheets and original documentation for the mentioned integrated circuits on the Internet and to study them in relation to this work.

The concepts presented are based on the concepts studied in selected earlier works (Darche 2000, 2002, 2003, 2004, 2012), which I recommend reading beforehand.

Philippe DARCHE
August 2020

Introduction

The two preceding volumes respectively addressed the hardware and software characteristics of the microprocessor. This last volume complements them by focusing on the software development chain and on development and testing tools. The final chapter in this book, which is also Part 2, presents structural changes in the first generations of microcomputers.

PART 1

The first part of this last volume is divided into two chapters. The first presents the software development chain, and the second illustrates the hardware and software tools used in development and testing.

1

Development Chain

This chapter is focused on the software development chain for a microprocessor-based system. The different steps and their associated tools, including the first, the assembler, are examined. The debugging and testing aspects have taken on greater importance as hardware and software have grown more complex and become embedded in systems. These software and hardware tools are then described in detail. At first conceived to be used at the printed circuit level, they must be integrated progressively. To conclude, assembly language, a first level symbolic language, will be surveyed. The reader who is interested in a specific processor should refer to documentation from the relevant manufacturer. In addition, we will not here make a distinction between microprocessor and microcontroller because, in the latter case, we will only examine the "computation" component, that is, the processor, without addressing the other two subsystems, which are Input/Output (I/O) and RAM/ROM (Random Access/Read-Only Memory), including programmable memory, as relates to the latter type.

NOTE.– An understanding of the concepts of data representation and arithmetic operations in a computer is assumed. Otherwise, *cf.* Darche (2000). This will also be the case for logical operators. On this subject, *cf.* Darche (2002).

1.1. Layers of languages, stages of development and tools

The processor uses two-state logic. Instruction codes and data are therefore expressed in binary. This is machine language. Manipulating such data is not humanly possible for a program longer than about a hundred lines. The first computers were programmed in this language, and the program (i.e. machine code and data) was entered in binary format using switches as input peripherals. Because of the difficulty of use, an additional layer of language closer to natural language (English) was therefore necessary. This language is called "Assembly Language," which is abbreviated in this work as AL. The term assembly emerged from the EDSAC project (*Electronic Delay Storage Automatic Calculator*, Wilkes 1950, 1951) defining the action of reading subprograms from a symbolic instruction tape (a letter), translating them into binary, and making them executable by the main program (modern functions of a link editor and loading program combined). This mechanism is attributed to David J. Wheeler (Wheeler 1950). Excluding machine-oriented language, assembly language is therefore the processor's base programming language. It is referred to as symbolic. This indicates that it manipulates symbolic information, instruction words, variables and labels primarily. A specific assembly language corresponds to a single Instruction Set Architecture (ISA, *cf.* § V1-3.5). Assembly instructions are referred to using a form of abbreviated writing called operation code (or opcode) mnemonic (*cf.* § V4-2.1). An opcode mnemonic is a symbolic representation of a machine code. The assembler is the computing tool that translates between the symbolic name and the binary value. The first assembler was written by Nathaniel Rochester for the IBM 701 (1954). Recall that instructions in machine language, also called machine code or sometimes hard code (Bell 1973), are a series of binary instructions that are read and executed by the microprocessor. Figure 1.1 shows an example of assembly with two lines of instructions for the 8086 microprocessor. The assembler translated the mnemonics into machine code, here expressed in base 16 for readability. This tool is specialized for a processor or a family of processors. There is no efficiency loss between assembly languages and binary because the translation is direct. This is not the case for High-Level (programming) Language (HLL) compilers, where one high-level instruction (i.e. statement) corresponds to a sequence of several instructions in assembly language.

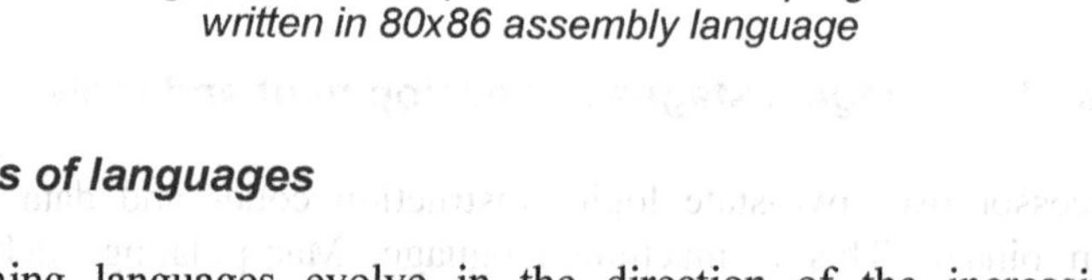

Figure 1.1. *Example of lines from a program written in 80x86 assembly language*

1.1.1. *Levels of languages*

Programming languages evolve in the direction of the increasing ease of programming by making it possible to manipulate higher-level abstractions, for example at the level of data structures or operators. Assembly language is downstream from high-level language, as shown in Figure 1.2. A language referred to as high level is as close as possible to human language. This is why machine and assembly languages are called Low-Level (programming) Languages[1] (LLL). To move from one level to a lower level, it is necessary to use a translator that will substitute a source-language instruction with a series of instructions belonging to the lower-level (target) language. Each instruction in machine language that is executed will give rise to a series of commands inside the microprocessor, or micro-operations. In a micro-programmed architecture, these commands are naturally called micro-instructions (this will be covered in a future book by the author on microprocessors). This microprogramming language belongs to the language level reserved for processor designers (not represented).

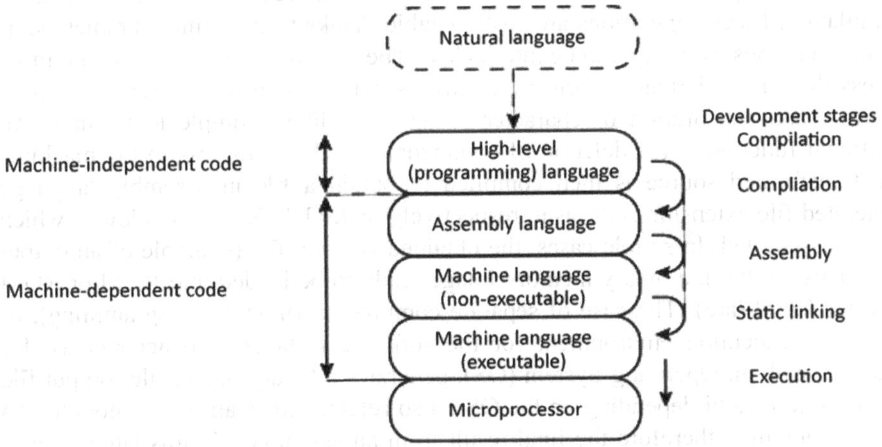

Figure 1.2. *Levels of languages in computing*

It should be noted that there are microcontrollers with on-board high-level language interpreters (*cf.* § V3-5.3).

1 High-level languages like C (Kernighan 1983) are sometimes classified by some authors (*cf.* Doyle (1985), for example) in the "low-level" category because they offer instructions close to the microprocessor such as sequential and combinatorial logic operators, and because they make it possible to manipulate variable addresses (concept of the pointer). This argument will not be adopted in this work because these languages provide high-level control structures and the idea of data typing, which is not the case for AL.

1.1.2. *Development stages*

Figure 1.3 presents the development chain for a computing application written in a compilable high-level language. It is also called the compilation chain or, more generally, the toolchain. A source program (or code) is written in a high-level language, for example in C (file extension .c) using a text editor. It is first precompiled (file extension .i for our example in the Microsoft environment or direct display on the standard output peripheral for UNIX). The precompiler is a preprocessor that will transform the source before delivering it to the compiler. It primarily provides functions for macro-expansion, macro(-definition) and file inclusion (*cf.* § 1.3.4). It also uses control structures to enable conditional compilation. Macro-operations are customizable thanks to its formal parameters, as are the functions, and they can be interlocked. The preprocessor is therefore no more or less than a manipulator of character strings with functions for search, deletion, insertion and substitution of character strings, just like a simple text editor. An additional function is the deletion of comments, which are useless to the machine. The transformed source is then compiled to obtain a file in assembly language (generated file extension .s or .asm, respectively under UNIX and Windows), which is then assembled. In simple cases, the obtained file is a file (example of an output file extension .hex), a binary memory image ready to be loaded into RAM or ROM (FW for FirmWare). The case of separate compilation (modular programming), or where the execution environment for the software is taken into account as, for example, with an Operating System (OS), generates a binary object file (output file extension .o or .obj depending on the OS), also referred to as an object module. An object program is therefore the final result from an assembler. In this latter case, it lacks code. This code corresponds to missing object modules. Once this code is available, the last step is called static linking. It is carried out by a linker or a link editor. This tool is generally automatically called by the compiler. It makes it possible to bring together all of the code forming the application based on a format that depends on the operating system. This missing code is presented in the form of independent object modules or compiled functions belonging to static libraries (file extension .a and .lib depending on the OS). This linker resolves address correspondence problems. The executable file is called a.out by default in UNIX; under Windows, it has a .exe extension or, formerly, a .com extension. This file can then be executed by the microprocessor (MPU for MicroProcessor Unit). To do so, it is loaded by the OS and then given execution control. The loader allocates memory, initializes the environment and can resolve address correspondence problems if necessary. It is also responsible for launching the program. To conclude, note that there are link editors/loaders.

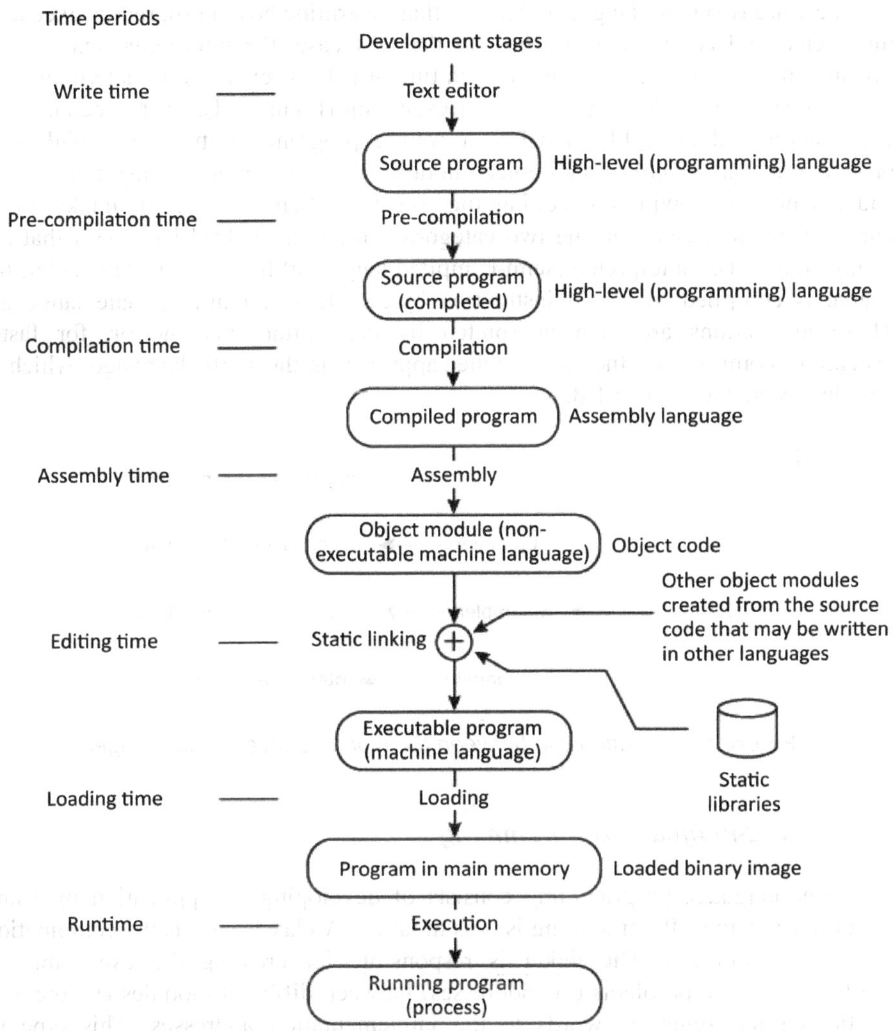

Figure 1.3. *Development chain for a program written in a compilable high-level language (here, C)*

There are two main language families that determine how a program is executed: interpreted and compiled languages. In the first case, the interpreter analyzes an instruction from the source program each time it is to be executed to determine how to do so, and this is done at the time of execution (Figure 1.4). In the second case, compilation and assembly translate a source program (written in a high-level programming language) into an object module. This takes place during compilation and assembly. Following link editing, the execution of an object program takes place during execution. Between the two categories, there are hybrid languages that are compiled and then interpreted (semi-compiled language) like Java. For the latter, the source is compiled to obtain instruction byte code in an intermediate language. These instructions are then interpreted by the virtual machine or, for faster execution, compiled on the fly. Another approach is the Forth language, which is both interpreted and compiled.

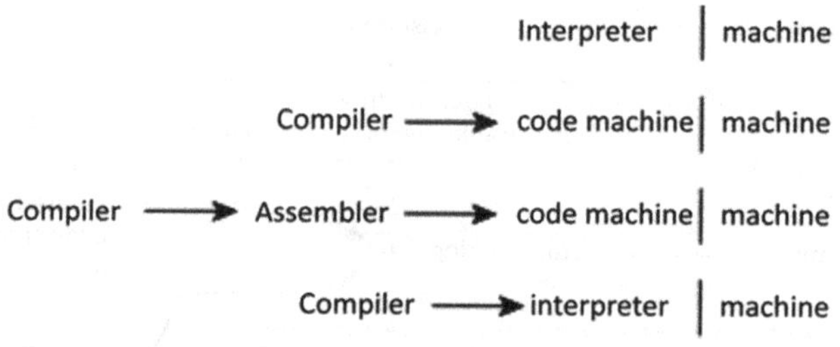

Figure 1.4. *Relationship between the type and levels of languages*

1.1.3. *Mixed-language programming*

Mixed-language programming consists of developing an application by using multiple languages. Programming is thus modular. A classic case is the combination of C, C++ and AL. The linker is responsible for creating the executable. It resolves reference problems (i.e. addresses) between different modules (Figure 1.5) by linking the symbolic words to the implementation addresses. This type of programming is complex and tends to generate errors. Each source is compiled with the specific language's compiler. At this stage, the tools become shared. Link editing brings together the various object modules to generate the application.

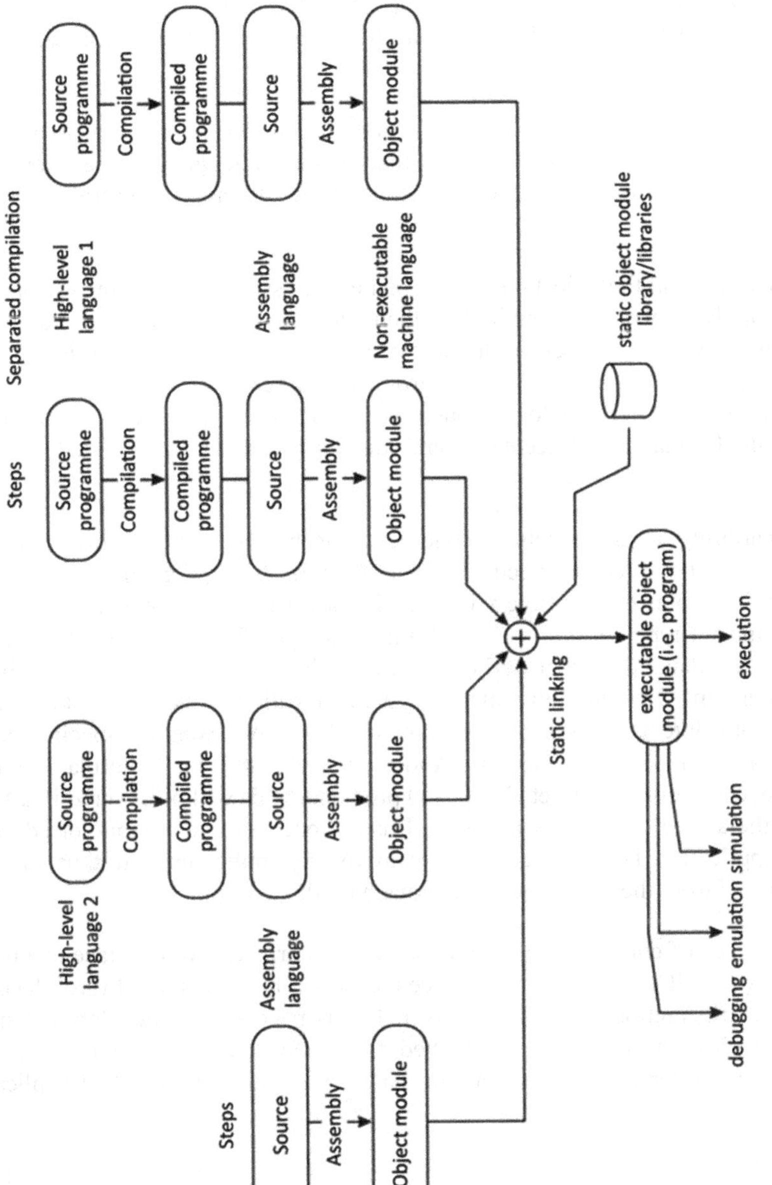

Figure 1.5. *Software development chain in a mixed-language environment and with separated compilation*

1.1.4. *Compatibility and software interfaces*

The concept of software compatibility is presented in § V4-3.3.2. It is relevant at three levels of the software development chain: source,[2] object and machine code (Figure 1.3).

Compatibility at the source code level today refers to high-level languages because it is better for obvious reasons than assembly language. These languages offer high-level software functionality (data type structures, control structures, paradigms, etc.).

Compatibility at the object code level makes it possible to distribute the program without supplying the source code. It is then necessary to carry out link editing (*cf.* § 1.2.2) with system libraries on the host computer, an example being libc in the C language. The specification takes over from the source code with, in addition, the definition of an object file format such as COFF or ELF (respectively Common Object File Format and Executable and Linkable Format, *cf.* § 1.2.2 for more details).

Compatibility at the machine code level or binary compatibility allows an application to be directly executed (ready-to-run). It requires the definition of symbolic data types (compatible with the high-level language declarations such as the long and int types in C, for example) and the specification of the alignment of structures and data. The other definitions are sections (code, data, stack and heap), their maximum sizes and the memory map (location), and the linker, which generates absolute addresses before knowing them. An example specification is BCS (Binary Compatibility Standard, Anderson *et al.* 1989) for Motorola's M88000 RISC (Reduced Instruction Set Computer) processor (this will be covered in a future book by the author on microprocessors). It is not necessary to compile or edit links for the application before execution (ready-to-run application). It thus provides independence from the programming language used.

These types of compatibility require standard interfaces. An interface is a bridge between two software layers or between a software and a hardware[3] layer. It provides an abstraction of the lower layer. It was necessary to develop a standard binary interface at the OS level (called the system interface), which is done symbolically, either at the programming language level with an API (Application

2 Or programs.

3 The hardware layers and their interfaces will be covered in a future book by the author.

Programming Interface) or at the binary level with an ABI (Application Binary Interface), as illustrated in Figure 1.6. A Hardware Abstraction Layer (HAL) makes it possible to detach the OS from the hardware implementation. ISA was described in § V1-3.5.

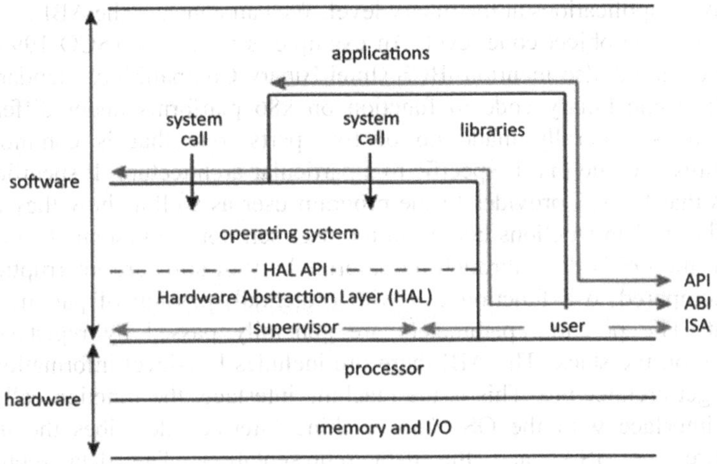

Figure 1.6. *API and ABI interfaces and Hardware Abstraction Layer. For a color version of this figure, see www.iste.co.uk/darche/microprocessor5.zip*

An API or application programming interface enables compatibility at the source level language in almost all cases for a high-level language (the C language, for example). It enables portability of applications at the source level (*cf.* § V4-3.2.3). It offers a set of services while masking the implementation details. The API is an abstraction for service calls. The invocation of these services is done in the form of standard or non-standard library functions. In the simplest case, the function is a simple wrapper that adapts/translates the high-level conventions into those of the lower-level. In the most complex cases, there are several system calls and the called function is processed. An example is SVID (System V Interface Definition) which defines the C language programming interface with UNIX System V (Novell 1995a, 1995b, 1995c). Another example is the IEEE 1003.1™-2013 – POSIX standard (IEEE 2013). This specification defines a set of executable functions, a standard library (library API) and a system API. A change in the API requires the application to be recompiled. The API belongs to the Abstract Machine Interface (AMI) located between applications and the OS. The latter also specifies the allowed instructions and the memory access model.

The ABI is a set of specifications to which the executable must conform so that it can be executed in a given execution environment. It defines a standard binary interface to be able to execute the compiled application. It is located between the applicable program and the OS, a library, or a set of I/O routines such as the BIOS (Basic Input/Output System, *cf.* § 4.2.2 in Darche (2003) and § 3.5.3). It enables portability of applications at the binary level. We can consider the ABI as equivalent to the API at the object code level. An example is System V (SCO 1996a, 1996b, 1997). We should also mention iBCS (Intel Binary Compatibility Standard), which allows the same binary code to function on x86 platforms under different Unix systems. It is generally made up of two parts, one that is common for all architectures and one that is specific to a particular architecture. It specifies a set of functions that the OS provides to the program user as well as how they should be called. The ABI instructions belong to the user set, not the system. The transfer of control to the OS is done through the intermediary of software interruption, which can be compared to a function call with (potential) passing of parameters under constraint. The potential parameters are generally passed by registers, or less commonly on the stack. The ABI therefore includes low-level information specific to the target architecture. This is the machine interface, the function call sequence and the interface with the OS. The machine interface describes the underlying architecture (i.e. ISA) and the data representation. The data representation specification provides its types with the format, order of storage (endianness; Cohen 1981, *cf.* § 2.6.2 in Darche (2012)) and associated alignments. The detail of the call sequence includes the register usage conventions, the stack frame layout convention and the passing of parameters (number, passing mode and type). The interface with the OS describes the exception interface, signal management, virtual addressing, process initialization (stack, registers, etc.) and debugging support. The executable and object file format are also specified (header, sections, etc.) as well as a library format. The ABI secondarily specifies the alphanumeric code used, for example, for character-based data control (\n, for example), or for the package data file. It provides information about loading the program and about potential dynamic linking. The ABI must support the API's libraries. Tools such as the compiler, the assembler and the linker, that is, the development chain, make reference to the ABI. An EABI (Embedded ABI) refers to the ABI version used in embedded systems. Its specific characteristics are the absence of a linker and modified memory management.

The interface between the OS and the hardware, the latter of which is optional, is called the Hardware Abstraction Layer. It is a software layer in the OS that abstracts the underlying hardware and is accessible via an API. An example is Windows. Drivers and hardware-specific software such as hot-swap management belong to this layer. It enables the addition of new hardware without having to change the OS's programs. Real-Time Operating Systems (RTOS) may not use a kernel (kernel-less approach), but instead a library linked to the application (library-based RTOS).

Using a standard API or ABI enables software compatibility (*cf.* § V4-3.3.2). The former enables application portability at the source-language level, and the latter does so at the binary level.

It should be noted that the idea of the machine is relative. As specified by Smith and Nair (2000), it depends on the point of view under consideration. As part of an operating system, the machine is the hardware support enabling execution, and the software interface is the ISA. As part of a process executing a program for a user, the machine is the OS, which supplies storage and memory as well as services such as I/O, and hardware support for execution is the ABI.

1.2. Fundamental software tools for development

The three primary tools required to develop an application are the assembler, the linker and the loader/launcher; there is also a tool called a disassembler. "Tuning and testing" will be addressed in the following chapter.

1.2.1. *Assembler*

The first assemblers were assembler-launchers (assemble-go system). They were responsible for loading subprograms written in assembly language into memory by translating binary instructions on the fly (i.e. a single program read), to call addresses[4] from the main program and to launch execution. In its modern form, it is a language translator. The first assemblers in modern form were SOAP (Symbolic Optimizer and Assembly Program) on the IBM 650 (Poley and Mitchell 1956) and then SAP[5] (SHARE Assembly Program) on the IBM 704 (Wegner 1976). A description of the latter can be found in Helwig (1975). Two representatives from the PC-Wintel[6] world are MASM (Microsoft Macro Assembler) and Turbo-Assembler® (tasm) from Borland.

The assembly of a source file will give rise to the creation of an output file named "listing" (the extension for the generated file is .lst), an object module, either absolute or relocatable, and a list of cross references (Figure 1.7). It should be noted that comments are deleted, which is handled by the precompiler (preprocessor) for a high-level language.

4 The term "address" refers to the logical address, a term belonging to the concept of Virtual Memory (VM, this will be covered in a future book by the author on storage). If there is no logical translation, it is a Physical Address (PA).

5 The original definition from IBM was *Symbolic Assembly Program*.

6 A contraction of the two names Windows and Intel.

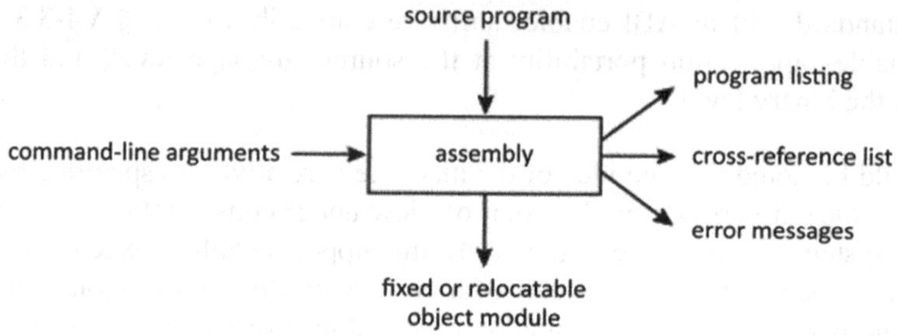

Figure 1.7. *Information flow for assembly*

Given that this is a symbolic language, translation is more than transliteration, hence the term assembly (Wegner 1968). Assembly can be broken down into two parts, analysis and synthesis. The latter can be divided into phases or steps (Figure 1.8). For a given language, syntax and semantics must be defined. Syntax concerns the formal aspects of a language. It manages the organization of instructions and data. Semantics concerns the meaning of instructions. Assembly will therefore begin with a lexical or lexicographic analysis. Its function is to pick out, from the data flow, a series of symbols called lexical or syntactical units, lexemes or tokens, from whence its other name, the scanner or tokenizer, or linear analysis, because the source is read from left to right, line by line. The typographic space is the base separator. There exist predefined lexical units in the language such as reserved words, operators, register names and separators (e.g. :, {, }, ;, etc.), as well as others defined by the user, identifiers[7] and numerical and alphanumeric literals (i.e. character strings). A second stage consists of determining to which family of units these lexemes belong. This is the sort phase carried out by the analysis. A data structure called a "symbol table" was initialized during these first two phases. This table will contain all of the elements to which the analyzer will add information such as their name, username, unit type such as identifier, constant (also called literal), language keyword or comment. These will also be initialized if applicable. An identifier is the name of a variable, constant, subprogram or label. It will then be elaborated during the following steps. From the point of view of implementation, these two functionally distinct steps are united (Wilhelm and Maurer 1994). This is followed by syntactic analysis, also called hierarchical analysis. It is carried out by the analyzer with the same descriptor (*parser*). It consists of verifying whether the source is respecting the grammar of the language. A grammar defines the rules for constructing sentences in the language, that is, lines of code containing declarations and instructions. One of the formal systems for

7 An identifier or username is a noun (*cf.* a character string) designating, among other things, a variable, a constant, a subprogram or a label.

describing a grammar is the Backus–Naur Form (BNF, Backus *et al.* 1960, 1963; Knuth 1964; ISO 1996). During this step, syntax errors are addressed. Semantic analysis is the penultimate step. It verifies, among other things, whether the variable format is coherent between registers and/or variables. Code generation is the final step. Code generation will involve symbolic evaluation. It associates an operation code or opcode with each instruction and, to each identifier, an address, when this correspondence is possible. If not, the linker stops associating to generate the final executable. The mode of assembly can be absolute or relative. This refers to the addressing mode (*cf.* § V4-1.2) used to reference the identifiers. Relative addressing enables address translation (relocation, *cf.* § V4-3.1.4), which makes it possible to install the code and the variables anywhere in memory without having to recalculate the addresses.

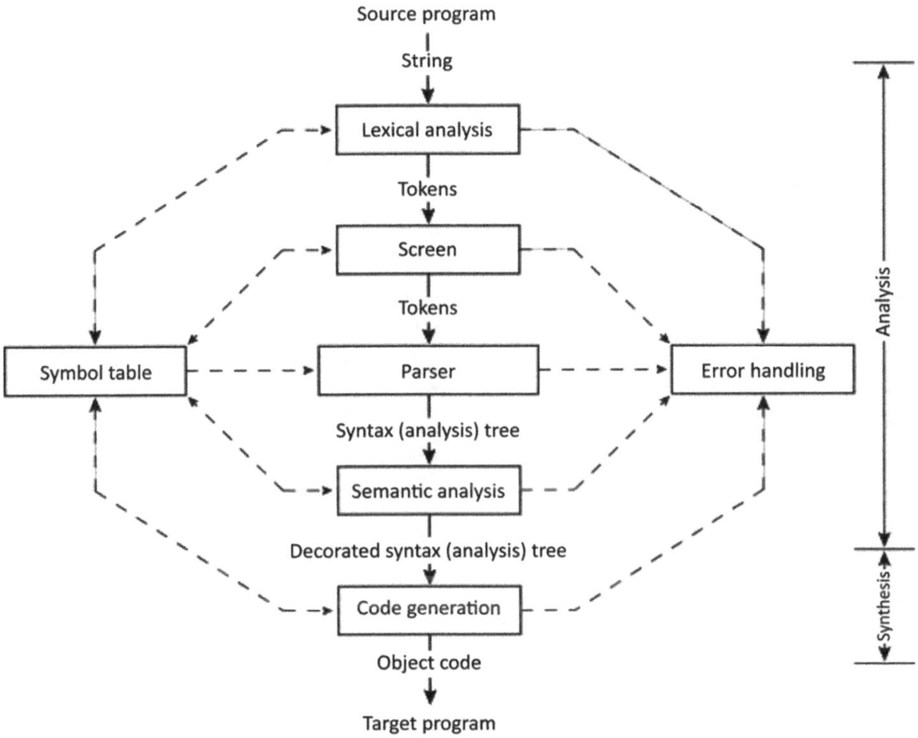

Figure 1.8. *Functional phases of assembly*

The structure of the symbol table will change depending on the step in the development chain. During assembly (assembly time), this data structure will contain, in its most complete form, the lexical units and, in less complete forms, only identifiers with their attributes. When included in an object module, it makes it possible to resolve references during linking time by connecting the identifier and the reference. During execution, this table, which can be optionally included, enables the debugger to know the names of the identifiers. Figure 1.9 shows an example of this kind of table, extracted from the listing file in Figure 1.10.

```
Turbo Assembler  Version 3.1        06/10/11 17:32:02       Page 2
Symbol Table


Symbol Name                        Type   Value

??DATE                             Text   "06/10/11"
??FILENAME                         Text   "div     "
??TIME                             Text   "17:32:02"
??VERSION                          Number 030A
@32BIT                             Text   0
@CODE                              Text   DGROUP
@CODESIZE                          Text   0
@CPU                               Text   0101H
@CURSEG                            Text   _TEXT
@DATA                              Text   DGROUP
@DATASIZE                          Text   0
@FILENAME                          Text   DIV
@INTERFACE                         Text   00H
@MODEL                             Text   1
@STACK                             Text   DGROUP
@WORDSIZE                          Text   2
BOUCLE                             Near   DGROUP:0012
D                                  Word   DGROUP:0002
DEBUT                              Near   DGROUP:0000
DIVO                               Near   DGROUP:0025
FIN                                Near   DGROUP:002C
FIN_MES                            Byte   DGROUP:0024
MESSAGE                            Byte   DGROUP:0008
N                                  Word   DGROUP:0000
OP                                 Near   DGROUP:000D
Q                                  Word   DGROUP:0004
R                                  Word   DGROUP:0006

Groups & Segments                  Bit Size Align  Combine Class

DGROUP                             Group
  STACK                            16  0100 Para   Stack   STACK
  _DATA                            16  0025 Word   Public  DATA
  _TEXT                            16  0030 Word   Public  CODE
```

Figure 1.9. *Example of a symbol table extracted from the listing file in Figure 1.10*

As with high-level languages, assembly can be conditional (CA for Conditional Assembly), which makes it possible to assemble only some parts of the program. An example use case is the potential to choose whether to include instructions from an application update. As with a compiler, the assembler, to carry out the translation, generally performs two passes through the source code. A mono-pass assembler directly translates symbolic references. But there are cases when this translation cannot be done, particularly when there is a forward reference for a variable or a label. There are therefore "two-pass," or even multi-pass assemblers. Assembly time is a function of the number of passes, of course.

The structure of the generated file, that is, the object module, can be broken down into a general information header, code and data, the area for useful address translation during a change of address, the global symbol table and the optional debugging information. The object module can be called absolute if the address of the program start was fixed and all of the identifiers' addresses are generated from it (this is the case in a mono-pass assembler). The memory dump therefore cannot be moved to main memory, contrary to a relocatable module. Modern assemblers generate this second kind of object module.

At the programmer's request, a listing file can be generated. The data line for this type of file (Figure 1.10) is typically divided into three areas called fields that are, from left to right, the line number (expressed in base-10 most of the time), the memory address (expressed in hexadecimal), the size, the variable name (optional) and the initialization value (optional), which corresponds to the assembly of the fourth zone, the source, as described in § 1.3.

The instruction line is typically divided into three areas called fields that are, from left to right, the line number (expressed in base-10 most of the time), the memory address (expressed in hexadecimal[8]), the machine code, then the source. A cross reference table can be constructed from the symbol table during assembly. For each symbol, whether internal to the program or external (i.e. public), there will be an entry in this table with its name, its value and the list of instructions it references, or even the number of the corresponding line. This can be seen in this file. This list facilitates debugging, among other things, and it can be used by a compiler.

8 This foundation enables a more compact representation of values compared to NBC (Natural Binary Code).

```
Turbo Assembler  Version 3.1         06/10/11 17:32:02        Page 1
div.asm

                                       ; nom du programme : div.asm
                                       ; division par additions successives
                                       ; une approche simple

                                              IDEAL
                                              DOSSEG
 0000                                         MODEL TINY
                                              P8086

 0000                                         STACK 100h

 0100                                         DATASEG
 0000   000C                           N      DW 12
 0002   000C                           D      DW 12
 0004   0000                           Q      DW 0
 0006   0000                           R      DW 0

 0008   44 69 76 69 73 69 6F+  message DB "Division par zéro impossible."
        6E 20 70 61 72 20 7A+
        82 72 6F 20 69 6D 70+
        6F 73 73 69 62 6C 65
 0024   24                     fin_mes DB "$"

 0025                                         CODESEG

 0000   B8 0000s               debut:  mov ax,@data
 0003   8E D8                          mov ds,ax
 0005   8E C0                          mov es,ax

 0007   8B 0E 0002r                    mov cx,[D]
 000B   E3 18                          jcxz div0

 000D                          op:
 000D   A1 0000r                       mov ax,[N]
 0010   33 DB                          xor bx,bx

 0012                          boucle:
 0012   43                             inc bx
 0013   8B D0                          mov dx,ax ; sauvegarde temporaire du reste
 0015   2B C1                          sub ax,cx
 0017   73 F9                          jnc boucle

 0019   4B                             dec bx
 001A   89 1E 0004r                    mov [Q],bx
 001E   89 16 0006r                    mov [R],dx
 0022   EB 08 90                       jmp fin

 0025                          div0:
 0025   BA 0008r                       mov dx,offset message
 0028   B4 09                          mov ah,9
 002A   CD 21                          int 21h

 002C                          fin:
 002C   B4 4C                          mov ah,4ch
 002E   CD 21                          int 21h
                                       END debut
```

Figure 1.10. *Example of a listing file generated by tasm (symbol table in Figure 1.9)*

There are different types of assemblers. A macro-assembler is an evolution from traditional assemblers that enables the creation of macro-instructions (*cf.* § 1.3.4), which, once substituted, add to the code. This makes it possible to simplifying how a program is written. A cross-assembler generates code for a different processor than the one executing the assembler, as opposed to a naive or resident assembler, also called a self-assembler. This is required when the target on which the application will be executed is not sufficiently powerful (i.e. MPU power and memory and storage capacities) to run the development tools. It is often used to develop embedded applications. A multiprocessor assembler is capable of generating code for microprocessors in a different family. There is also something called a meta-assembler, a scientific curiosity that makes it possible to write an assembler as they were initially described (Ferguson 1966). A High-Level Assembler (HLA) approaches high-level languages by enabling variable types and using control structures and macro-instructions. One of the first was described by Wirth (1968) with his PL360 language for the IBM System/360. A modern version of this type of language is HLA, proposed by Hyde (2010). Some high-level language compilers, such as the one for C, make it possible to insert lines of assembly language. This is referred to as an inline assembler. This is useful for directly accessing the MPU's instruction set or for calls to the OS. The downside is the loss of portability since this language is specific to an architecture or a specific MPU. Finally, we can mention the patch assembler, which is used in debuggers to modify application code in real-time to correct an error and test it immediately.

1.2.2. *Linker*

During the last step before obtaining the program executable, the linker (linkage editor) or link editor provides address translation functions (relocation, *cf.* § V4-3.1.4) and symbolic resolution. It selects the implementation address for the code and the variables as a function of the memory model, for example. It defines a size as an assembler and a start address (TOS for Top Of Stack) for the stack. If all of the modules are presented, it resolves the address correspondence if necessary. Generally, from the object modules and static and dynamic libraries (or DLL, for Dynamic Link Library), that is, shared between several applications, and following the arguments entered on the command line, it generates a module, either an executable or a relocatable, as illustrated in Figure 1.11. The executable file has an extension of .com, .exe (Windows environment), or an execution permission[9] (x for UNIX), or is a file for programming an EPROM (file extension .hex, for example). The second case, in which a new relocatable object module is generated, is due to the fact that the linker does not continue to the end, that is, that there remain unresolved references because, for example, of a missing library. Symbolic information generated optionally during assembly can be supplied as well for debugging at the executable source level (traditional file extension .map). This information can be related to sections (number, type, start address and size) and

9 It is no longer a question of having run permissions because a program can be executed or interpreted.

identifiers (name, type, address, allocation class, etc.). A control file (LCF for Linker Control File), also called a linker script file, contains the instruction to organize these sections and to define memory locations. As during compilation and assembly, there exists the notion of a pass or passes.

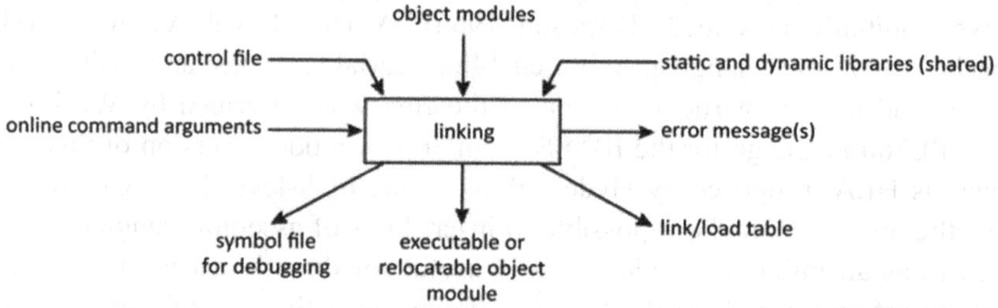

Figure 1.11. *I/O for link editing*

A library is a collection of general-use object modules (i.e. compiled functions) collected by domain or subject (e.g. the standard language library, a graphics library, etc.) that is intended for reuse. This idea of a library is more present in high-level languages like C with, for example, its standard library, than in AL. That said, there was an attempt to offer a standard library by the University of California Riverside named the UCR Standard Assembly Language Library, for use with the 80x86 family. There are three types: static, dynamic and import.

A static library or archive contains relocatable object modules. At link time, the object module's code is included in the output file, which is its major disadvantage for large applications.

A dynamic link library (file extension .dll on Windows), also referred to as a shared library (hence its file extension .so for shared object under UNIX) is a library of compile functions shared between applications that, instead of being linked definitively to the application code during static linking, are stored in memory upon execution. A distinction can therefore be made between dynamic linking and dynamic loading, as illustrated in Figure 1.12. Dynamic linking takes place when the application is launched. Dynamic loading takes place on demand when called, therefore under the program's control. Linking is delayed at execution (runtime) in both cases and is therefore done dynamically. Postponing the loading of a shared library until it is called allows for an increase in available memory space and time saving for the initial launching of the application. This latter case is called lazy linking. This approach favors modularity and code re-use. Under Windows OS, we find libraries in the form of files with extensions such as .dll, .ocx, .cpl or .drv. Under Unix, libraries carry the extension .so (shared object).

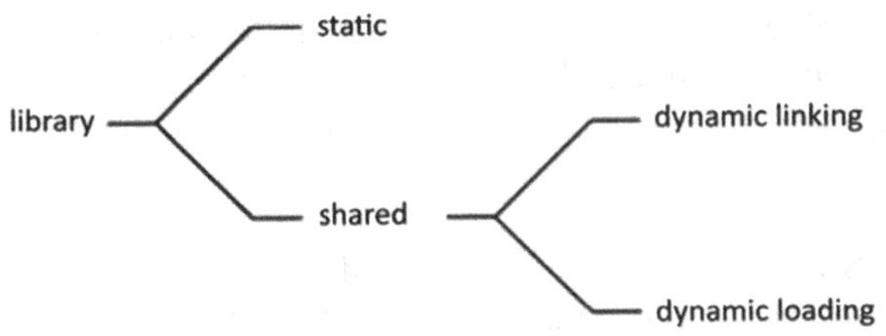

Figure 1.12. *Types of libraries*

An import library is a hybrid that enables management of loading and use of the shared library. Under Windows, it occurs in the form of a static library with the same name as the dynamic library and is statically linked to the application.

The generated file format is generally dependent on the OS. The format a.out, originally a UNIX format, is made up of a header, the code or text section, the data section and other sections. The COFF format, which originated on UNIX System V, is a format for executable files, objects and libraries introduced in UNIX System V to replace the a.out format. It was adopted by Microsoft. This format has itself been replaced on most systems by ELF, with the notable exceptions of the DJGPP compiler for object files and of Microsoft Windows, whose Portable Executable (PE) format is a modified version. This is a much more complex format because it can be composed of an arbitrary number of sections, such as, for example, under Linux, with .text for instructions, .data for initialized variables, and .bss for uninitialized variables, .rodata for constants, .symtab for the symbol table, etc. There are other formats like those in the PC (Personal Computer) world. We can cite Intel's OMF (Object Module Format) and Microsoft's PE format, which is derived from COFF. The latter is the executable file format (file extensions: .exe and .ocx – ActiveX and OLE) and dynamic files for 32-bit Windows operating systems.

To manage a library, there is an archiver/librarian. The main functions of this archiver are insertion at any position, extraction, displacement, replacement and deletion of an object module in a file named library in the standard format (COFF, for example). It also lists its contents. Under Unix, we can mention examples of archivers such as `ar` and `ranlib`.

For the sake of completeness, the linker uses a binding form of linking.

1.2.3. *Loader/launcher*

The loader transfers a program that is to be associated with a process from mass storage to main memory. In the early days of programming, it was associated with the assembler (assemble-go-loader) and the linker. In the simplest case, it consists of a simple copy operation. Historically, loading was done at a fixed address. Today, for the sake of memory management, the load address is variable. Other functions can be added such as memory space allocation.

Loading the OS is a special case. It is carried out by the startup program called BIOS (*cf.* § 4.2.2 in Darche (2003) and § 3.5.3). This firmware contains a program called a primary boot loader that is responsible for loading and launching the secondary boot loader at sector 1 of track 0 (cylinder 0 head 0, more precisely) (MBR or Master Boot Record) or a launcher such as grub for Linux. It should be noted that there also exists network launching, or boot by LAN (Local Area Network).

Putting this software into ROM makes it impossible to add new support for peripherals. This is why BIOS has evolved towards UEFI (Unified Extensible Firmware Interface, *cf.* § 3.5.3), an initiative begun by Intel.

1.2.4. *Disassembler*

This section cannot conclude without addressing the disassembler, which is the flip-side of the assembler. The idea is to be able to reconstruct the source from the object module. This can be necessary, for example, if the source was lost inadvertently or for the purpose of reverse engineering. It analyzes the machine code to retrieve the mnemonics. But recovering data structures and comments is problematic. Effectively, the tool has no information to recover symbolic names. This can be done by integrating information during assembly that is useful for disassembly and debugging, such as a symbol table or the compiler name. To do so, compilation and assembly must be done with the necessary options to include this information. We can distinguish between traditional disassembly of a program's machine code and the kind that comes from traces (*cf.* § 2.2).

1.3. Assembly language

Assembly language is a symbolic representation of machine-executable binary instructions. Table 1.1 shows a comparison between high-level (HLL) and low-level (programming) languages with regard to size and development time. Jones (1998) also compares this language to COBOL (COmmon Business Oriented Language) by adding the criteria of function points (Albrecht 1979). We see that assembly language provides more code and that the development time is greater, at least double in relation to C. On the other hand, AL provides better productivity because the generated code is closer to the machine. But application maintenance is more difficult than with HLL.

Languages	Size (in lines of code)	Development effort (in man/months)
Assembly	10,000	40
Macro-Assembly	6,666	28
C	5,000	22
COBOL	3,333	16
Pascal	2,500	13
Ada	2,222	12
Basic	2,000	10

Table 1.1. *Language comparison (based on Jones (1986))*

Why use assembly language? Originally, it was used to avoid the need to directly work with binary code. By providing, among other things, symbolic notation for identifiers and instructions, assembly language offers an ease of use that machine language, which works with 1s and 0s, cannot. AL is the generable and readable version of machine language. Symbolic names enable accelerated development and decrease the number of programming errors. It makes it possible to eliminate syntactic errors (misuse, missing instruction, bad machine code, etc.), which will be detected during the assembly phase (*cf.* § 1.2.1). Commands facilitate programming. With the invention of high-level languages, assembly language has been progressively restricted to niches where the increase in execution speed and memory space is a priority. Its instructions are those from the MPU's instruction set. A program written in assembly language is therefore more efficient in terms of speed and memory usage than a program written in a high-level language. Examples include time-critical software functions such as schedulers and low-level I/O drivers for a real-time operating system or a loader. Originally, MPU evaluation kits (*cf.* § 2.1.1) only contained a few kibibytes (KiB) of memory! Embedded and on-board systems require even more economic use of memory space. Working near the processor makes it possible to optimize memory space and achieve compact code. Microcontrollers, components that even now contain limited memory, are still programmed using AL. Furthermore, a microprocessor that is not in widespread use may not have a compiler for a given language. AL is the only recourse in this case.

Assembly language is of pedagogical interest to help students understand in detail the operation of the MPU, as well as some of the mechanisms of high-level languages. Without detailed knowledge of MPU programming, it is not possible to understand the internal operation of this component. Superscalar architecture (this will be covered in a future book by the author on microprocessors), for example, cannot be understood without knowledge of the idea of branching or register addressing. It can also provide an introduction to learning about architecture. From the software perspective, it is also possible to study concepts such as function calls and passing parameters, control structures in high-level languages, and, thus, possibly, to learn how to program more efficiently through knowledge of how the compiler will translate the program's source code. Operating system mechanisms such as system calls are easily explained with AL. It is also possible, in order to not derail the student, to program in the style of a high-level language (Crookes 1983).

But today, operating systems include several tens of millions of lines of code. Programming in assembly language is no longer humanly possible. On the other hand, it can also be used to create specific interfaces not provided by the high-level language. For example, it is possible to use functions written in another language using an interface written in AL. High-level languages or an operating system may not provide the primitives necessary to precisely manage an "exotic" and therefore non-standard programmable component. Calls can be made specifically, for example, with the help of a specialized instruction or by using a register. This is particularly true for I/O. It is also possible to add primitives from an operating system in a high-level language. It also enables the production of specific interfaces not provided by the language, such as a peripheral driver. AL provides complete access to the MPU's instruction set, for example those that process multimedia (*cf.* § V4-2.7.1). This allows a programmer using a high-level language to access hardware using the inline assembler more efficiently. Embedded systems are another area. These are generally programmed using several types of language, one or more high-level languages for high-level applications, and AL for interactions between the operating system and the hardware (storage and I/O). Finally, the link editor, such as the one in Linux, can call a module written in AL to generate an executable.

It is important to note that memory is less and less expensive. Compilers for high-level languages (C, C++, Pascal, Ada, etc.) generate increasingly optimized code when evaluated based on criteria including runtime and storage/memory use. Furthermore, increasingly powerful microprocessors can undergo a loss of power because of non-optimized code generated by a compiler. Generation of code in assembly language can only be automated. Another major inconvenience is that this language is not portable or provides limited portability, limited to a family of MPUs. Inserting code inline in AL therefore eliminates the high-level language's portability (*cf.* § V4-3.2.3). A program therefore has to be rewritten for each microprocessor (instruction or equivalent block of instructions). Memory alignment (*cf.* § 2.6.1 in Darche (2012)), if the insertion is done directly to the machine code, can lead to a decrease in performance (*cf.* § V4-3.4). Upward compatibility (*cf.* § V4-3.3.3) of code, as is the case with the x86 family, is useful, but requires the designer to make some of the architecture inalterable or to add complexity. Furthermore, this language is more difficult to learn compared to a high-level language. Programs are more difficult and take longer to write, read, understand, debug and maintain.

From the architecture perspective, modern approaches such as VLIW (Very Long Instruction Word, this will be covered in a future book by the author on microprocessors) or speculative execution make it difficult to use AL because their complexity is offloaded to the compiler. Even if AL programming is still possible, it does not enable the optimal use of the underlying architecture.

1.3.1. *Software development methodology*

As for any language, an analysis phase for the problem to be addressed is necessary before any program is written. A processing algorithm is described textually with the help of a pseudo-language or designed using a flow chart (the use of which is in the process of dying out), which is a logical representation of the algorithm followed by a flow chart specific to the machine. Then, the program can be written using a text editor, at first in text mode (originally, now abandoned), initially a line editor and now a full-page editor (modern form).

1.3.2. *Standardization of assembly language*

Every designer of a processing unit or microprocessor proposes their own names for instructions and modes for addressing, which introduced confusion or even errors during design and programming. Furthermore, assembly language, because it is specialized for a processor or family of processors, is not portable. Attempts to propose a generic assembly language, for example by Nicoud (1976), proved vain. A standard was issued to attempt to resolve the problem. This was the IEEE 694-1985 standard (IEEE 1985).

1.3.3. *Structure of a program*

Regardless of the microprocessor and the assembler, programs in assembly language are similar from the structural and syntactic perspective. This section also presents a generic model for a program. A program is traditionally broken down into three sections: assembly commands, data and instructions, as shown in Figures 1.9/1.10 and 1.13. The separator is the space or the tab symbol, which added a predefined number of spaces.

```
/*
 * test_led_switch.asm
 *
 *  Created: 07/05/2014 16:48:49
 *  Author: Atmel
 */

;***** STK500 LEDS and SWITCH demonstration
.include <8515def.inc>
.def Temp =r16 ; Temporary register
.def Delay =r17 ; Delay variable 1
.def Delay2 =r18 ; Delay variable 2

;***** Initialization
RESET:
      ser Temp
      out DDRB,Temp ; Set PORTB to output

;**** Test input/output
LOOP:
      out PORTB,temp ; Update LEDS
      sbis PIND,0x00 ; If (Port D, pin0 == 0)
      inc Temp ; then count LEDS one down
      sbis PIND,0x01 ; If (Port D, pin1 == 0)

      dec Temp ; then count LEDS one up
      sbis PIND,0x02 ; If (Port D, pin2 == 0)
      ror Temp ; then rotate LEDS one right
      sbis PIND,0x03 ; If (Port D, pin3 == 0)
      rol Temp ; then rotate LEDS one left
      sbis PIND,0x04 ; If (Port D, pin4 == 0)
      com Temp ; then invert all LEDS
      sbis PIND,0x05 ; If (Port D, pin5 == 0)
      neg Temp ; then invert all LEDS and add 1
      sbis PIND,0x06 ; If (Port D, pin6 == 0)
      swap Temp ; then swap nibbles of LEDS

;**** Now wait a while to make LED changes visible.
DLY:
      dec Delay
      brne DLY
      dec Delay2
      brne DLY
      rjmp LOOP ; Repeat loop forever
```

Figure 1.13. *Example of a source file for a microcontroller from the Atmel AVR family*

An assembly directive or pseudo-operation (a word taken from the SAP assembler) can have two functions. The first is to direct the assembly or link editing by providing non-deductible data. This can include, for example, the target operating system definition or a specific syntax to be used. The second concerns data structure and initialization (data directives) and the structure of instructions. This includes the definition of variables and memory areas. A directive is therefore not translated into instructions for the machine. It is instead an instruction for directing assembly. It is also called a pseudo-instruction. Here are some typical examples. The ORG directive makes it possible to specify the load address, also called the assembly address. The end of the program is indicated with the END directive. EQU makes it possible to define a symbolic constant. This directive assigns a numerical value to a symbolic name (similar to but not exactly the same as #define in C with a parameterless macro). It is also possible to define variables, with or without initialization. To do so, memory space must be reserved (using the RMB directive for Reserve Memory Byte) and initialized. A macro-instruction (*cf.* § 1.3.4) is predefined using the MACRO directive. The directive is sometimes preceded by a period. Some AL may provide file inclusion (.inc directive, for example), similar to what C permits with its #include directive.

The data section enables the definition of variables with or without names and the specification of their format, with initialization if necessary. Normally, it begins with DATA. The label field enables the definition of variable names. The symbolic name will be assigned the current address value during assembly. An alignment directive may also be present. The syntax of a line is therefore as follows:

```
[label]     format [value]   [; comment]
VAR1        DB 4             ; variable definition
                             ; initialized to 4
```

Unlike a high-level language, AL does not offer data typing as a standard feature, but it is possible to define a format (i.e. the number of bits or bytes). The numerical values should be expressed with a specified or implicit base, generally base-10. By using prefixes[10] or suffixes, whether normalized or not, B for Binary, D for Decimal, H for Hexadecimal and Q for Octal. It is therefore implied, since originally there were only natural numbers. The type, a whole number or integer, is only revealed by the indicators. With the evolution of hardware, this language today offers data formats including both natural numbers and integers, floating and fixed-point numbers, character (strings), and binary (bit-string format). For character strings, the value is enclosed by double quotes and may include a directive that indicates the encoding (.ascii or .asciiz, for example). An example is MASM from Microsoft.

Under UNIX, the instructions section is called the text section. A line of instruction in a source file (Figures 1.9/1.10 and 1.13) are typically divided into four fields, which are, from left to right, the label, the instruction mnemonic, operator field, or operation with, if necessary, an operand field and, to conclude, the

10 For reference, the C language uses the prefixes 0x for hexadecimal, 0 for bytes and sometimes 0b for a binary pattern.

comments. We can therefore define its syntax and an example for the x86 family (Intel):

```
[label] [instruction] [operand[,operand]]     [; comment]
            mov ax,0
            mov cx,10
loop_begin:
            add ax,1             ; beginning of the loop
            loop loop_begin  ; end of the loop
```

The location or label field is filled with a character string, which makes it possible to find an instruction line, that is, a determined point in the code, for example, the destination of a jump. The label makes it possible to automatically calculate (i.e. with the assembler) the address of a jump, for example. To do so, the assembler maintains an up-to-date location counter that is incremented each time it passes to the next instruction by a value that depends on the length of the current instruction. This counter is reset to zero when the section changes, or it initialized to a defined value by a directive (ORG, for example). The definition of a label should begin at the beginning of a line and, potentially, should be terminated with the character ":". There is generally a label indicating the beginning of the program to the link editor (e.g. the label _start), which can also be called the entry point. The code field is filled with a mnemonic. The operand field contains between 0 and n-1 values or references depending on the addressing mode and the architecture. It is important to distinguish between an operand for an operation (mathematical definition) and, in assembly language, operands for an instruction, which could either be a "traditional" operand or a result. Depending on the style, the destination operand can be to the left,[11] as is the case for high-level languages (Intel or MPU DLX (DeLuXe) style), or to the right, since allocation is done from left to right (AT&T style, used in UNIX BSD (Berkeley Software Distribution)). The comment field is very important in AL source code, because the language is complex, and the source code is hard to read compared to high-level languages. It begins with the comment delimiter, which is the ";" character.

The address mode for the operand is indicated by the syntax. Square brackets typically indicate whether the mode is indexed or indirect. Table 1.2 specifies the standard conventions. The current address (location counter reference) is indicated by an asterisk (typographical symbol *) in the standard, or otherwise by the $ symbol. This makes it possible to calculate a jump address, for example.

11 This is called reverse ordering.

Address space operators	Symbols
literal value	#
direct address	/
register	.
relative address	$
direct page, base page, page zero	!
indexed or direct	[]

Table 1.2. *Address space operators in the IEEE 694-1985 standard*

In the IEEE 694-1985 standard, the working format for the instruction is indicated by adding, after a period, the mnemonic or, less commonly, its operand, following the standard in Table 1.3. A character string is initialized by putting the text in quotation marks. Typing for floating-point numbers is done in the same way, with FS for Short Floating Point, FL for Long Floating Point, FX for Extended Floating Point and FP for Packed decimal Floating Point.

Format (bit)	Mnemonic suffix	Equivalent
1	I, I1	
x	Ix, x natural number	
8	B	I8
16	S	I16, B2
32	L	I32, B4, S2
64	Q	I64, B8, S4, L2

Table 1.3. *Standardized format suffixes*

A comment, which is optional but strongly advised, enables documentation of a program to facilitate reading and maintenance of the source code. It is not taken into account during assembly. It begins with an agreed-upon character, in this case the semicolon. This beginning marker varies between manufacturers. It can also be @, #, *, or as is the case in C and C++., /* */ and //.

Finally, with regard to case, Intel, Microsoft (ASM86), Motorola, Zilog, Arm®, SGS and TI use capital letters in code. Keil follows the same convention, except for registers! Microchip, MIPS, IBM (Power architecture), Intel (IA-64 architecture) and HP use lower-case letters in code. Atmel uses capital letters for documentation and lower-case letters in programming. Inmos puts mnemonics and registers in lower case in both documentation and programming.

1.3.4. *Macro-instructions*

The concept of a macro-instruction emerged from the EDSAC project in the term open subroutine (Wilkes *et al.* 1951). It was extended to high-level languages several years later (McIlroy 1960). A macro-instruction (for a character string substitution) or, in abbreviated form, a macro, is a symbolic name that can be substituted by a character string. It is also called a pseudo-instruction or synthetic instruction. This text can include directives, mnemonics or data declarations. To define one, a macro-definition must be used. They help avoid repetition in code, but are not functions, because the code is not factored. A macro enables the definition of new instructions or the use of higher-level constructions, close to those of high-level languages. In this way, it simplifies syntax and programming. A macro is customizable, which increases its expressive power. During assembly, formal parameters are replaced by actual parameters. A library of macro-instructions facilitates reuse. Here, we are approaching the syntax of a High-Level Language (HLL). A macro-assembler is an assembler that uses macro-instructions. During an assembly pass, the macro-instruction is expanded. Figure 1.14 shows an example macro followed by its expansion. Each language provides predefined macros to facilitate programming.

```
;  macro example
;  name: PUSH_REGS
;  function: stacking two registers
;  two parametres; register names

.macro PUST_REGS
     push @0
     push @1
.endmacros

utilisation:
     ldi          R18,0x00
     ldi          R17,0x02
     PUSH_REGS    R18, R17

expansion:
     ldi          R18,0x00
     ldi          R17,0x02
     push         R18       ;macro code
     push         R17       ;macro code
```

Figure 1.14. *Example of a program with macro-instruction and then expansion*

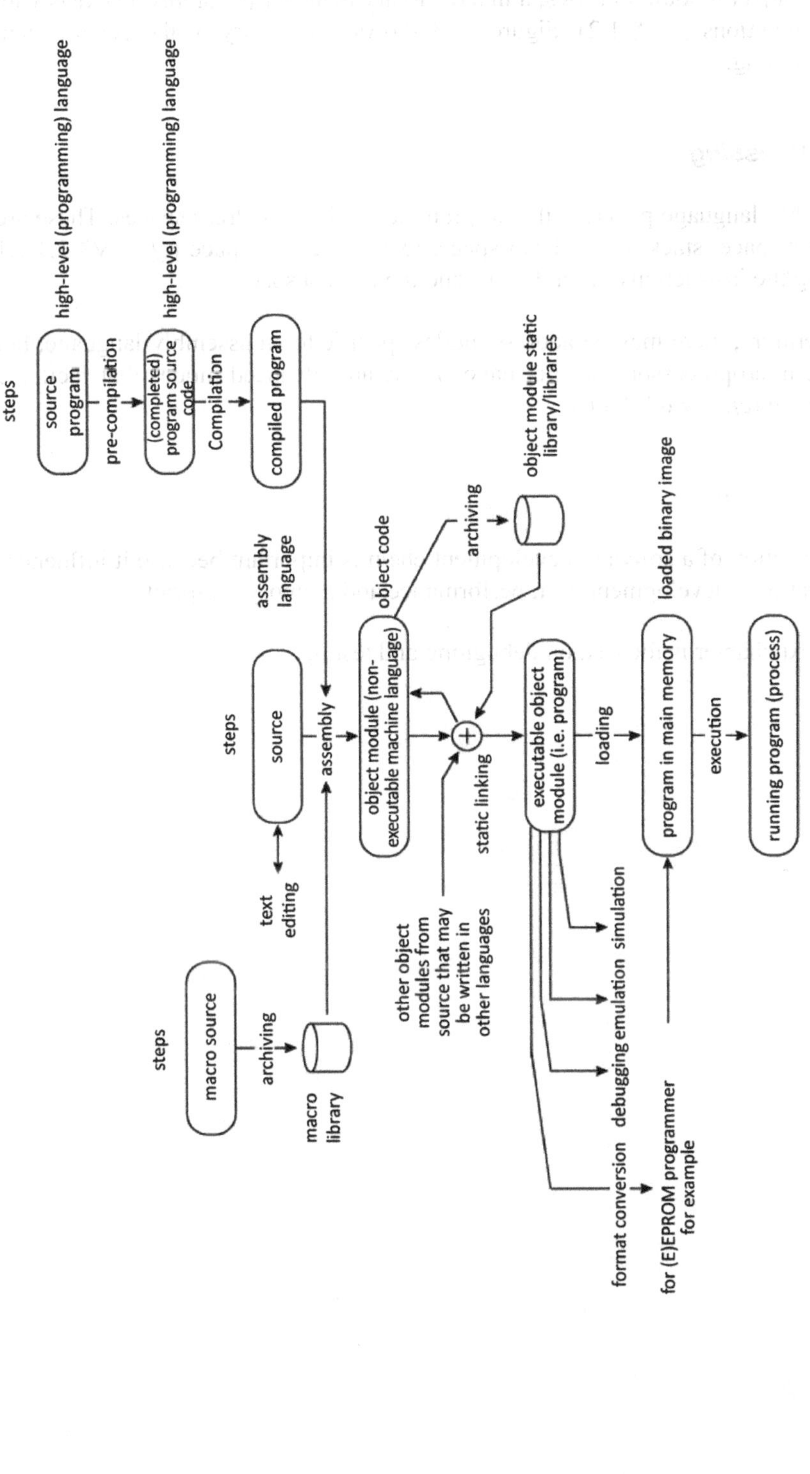

Figure 1.15. *Development chain with macro-management*

As for object module libraries, a macro-library manager (or archiver) exists with the same functions (*cf.* § 1.2). Figure 1.15 shows a summary of the development chain and tuning.

1.3.5. *Addressing*

Assembly language provides the programmer with six address spaces. These are the register space, stack space, heap space, text space, I/O space (*cf.* § V3-2.1.1.1 concerning the instructions `in` and `out`) and the control space.

Furthermore, there may be address modes specific to an assembly language, but which the microprocessor does not have. These are advanced modes that facilitate programming (*cf.* § V4-1.2.4.6).

1.4. Conclusion

The selection of a software development chain is important because it influences the application's development cost, performance and memory footprint.

The next chapter is focused on debugging and testing.

2

Debugging and Testing

This chapter focuses on development, debugging and testing for MPUs (MicroProcessor Units) for evaluating the performance of components and microcontroller programmers. It begins by describing the first electronic boards developed by designers. Debugging is then addressed from the hardware and software perspectives. The chapter concludes by addressing the testing aspect.

2.1. Hardware support

Various systems make it possible to evaluate an MPU. Moreover, programmers have been developed for microcontrollers.

2.1.1. *Generic electronic boards*

Before designing an application, a developer can take advantage of a range of generic electronic boards initially provided by microprocessor manufacturers, then by third-party companies. They make it possible to evaluate components and to quickly produce demonstration programs. In increasing order of performance, and with the caveat that there are subtle differences between them, there are starter kits, evaluation boards and development boards. The first, because of their very low cost, are made up only of a microprocessor and a minimal set of connectors. Evaluation boards have a more complete set of connectors and additional peripheral components, such as a more advanced initialization circuit or components that belong to an associated family of circuits. Development boards are evaluation boards to which the necessary components have been added for developing simple prototypes, that is, additional components such as memory, for example, or I/O (Input/Output) controllers and a debugging and programming interface (*cf.* § 2.2.5). Subsequently, the developer can invest in a more powerful development environment with full knowledge of the relevant factors. Demonstration boards are representative realizations implementing the component. These boards are therefore an inexpensive way to get to know a product and to allow the client to determine if it is a good choice for them. Economic requirements are therefore essential. Indeed, the price must be sufficiently attractive for the client to be able to try out the component at the lowest possible cost and then buy a complete system or build their own application, if necessary. With the Internet of Things (IoT), System on Modules (SoM) appeared, which combine evaluation sensors with processing and communication electronics on a small Printed Circuit Board (PCB).

There are several evaluation boards that were the key to the success of 8-bit microprocessors. These include the KIM1 board from MOS Technology and Rockwell; the Versatile Input Monitor (VIM), renamed the SIM Model 1 (SYM-1) from Synertek Systems Corp.; the Motorola MEK6800D2; the Intel SDK85 (Figure 2.1); and the AIM 65 from Rockwell International. The first four have between 1 and 4 kibibytes of RAM (Random Access Memory), a monitor of a few kibibytes of ROM (Read-Only Memory, *cf.* § 2.2.4), a 16-key input for hexadecimal numbers with additional 8 function keys, 6 seven-segment LED (Light-Emitting Diode) displays, a connector providing access to all signals on the bus and a soldering or wire-wrap prototyping area, if there is one. The AIM-65 board, which was more advanced, possesses a 20-character display and a 20-column dot matrix printer with 5 x 7 dot resolution. Some had a connection to a cassette tape, which made it possible to load and save a user program in the form of a signal on an audio tape (*cf.* § 7.2.2 in Darche (2003))!

The current trend is towards the elimination of the aforementioned peripherals, undoubtedly for reasons of cost and development flexibility, as well as because of the integration of development-facilitating functionality. The board is now made up of the microprocessor, RAM, flash EEPROM (Electrically Erasable Programmable ROM) and a serial communication interface that is usually RS-232 compliant (*cf.* § 8.2.2 in Darche (2003)) for historical reasons and because it is available on all microcomputers, if necessary, by using a USB (Universal Serial Bus) adapter. The board communicates with the user through an RS-232 asynchronous serial connection (initially) and a computer, even if the aforementioned peripherals are later added based on requirements (Figure 2.2). The user therefore has access to it in addition to the computer's resources for development. Applications are developed using a cross-compiler or cross-assembler (*cf.* § 1.2.1) and then simulated. They can then be downloaded for *in situ* debugging. The development computer can store source code, executables and memory dumps.

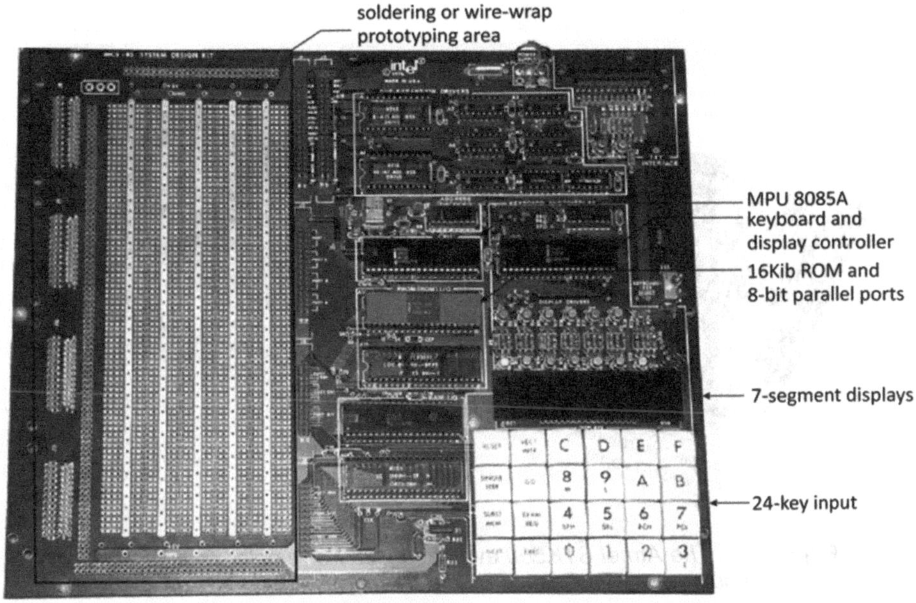

Figure 2.1. *The Intel SDK85 evaluation kit. For a color version of this figure, see www.iste.co.uk/darche/microprocessor5.zip*

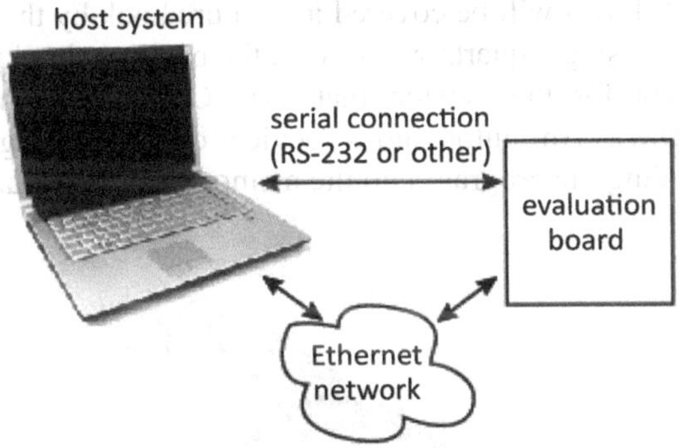

Figure 2.2. *Communication between host and target systems. For a color version of this figure, see www.iste.co.uk/darche/microprocessor5.zip*

Figure 2.3 shows the STK500 evaluation board for Atmel microcontrollers. For this type of component, these boards generally integrate simple input peripherals such as push buttons or outputs such as LEDs; in this example, there are eight of each.

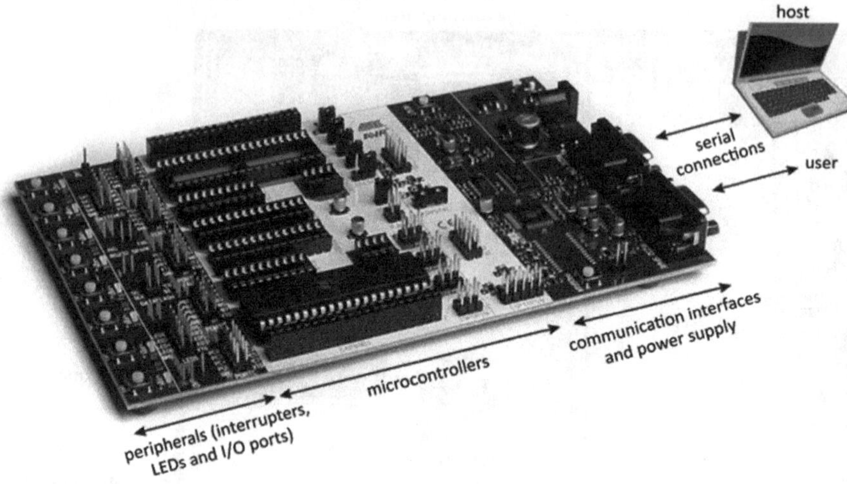

Figure 2.3. *The Atmel STK500 evaluation kit. For a color version of this figure, see www.iste.co.uk/darche/microprocessor5.zip*

2.1.2. *Programmers*

The first microcontrollers had UV-EPROM memory (Ultra-Violet Erasable Programmable ROM, this will be covered in a future book by the author on storage) that could be erased using a quartz window on the package that let through UV light (Figure 2.4), except for the version that used OTP (One-Time Programmable) EPROM (OTPROM). An autonomous device called a programmer that was responsible for loading the program into the memory was necessary.

Figure 2.4. *Microcontroller with integrated EPROM seen through the package's quartz window. For a color version of this figure, see www.iste.co.uk/darche/microprocessor5.zip*

Today, ROM is usually Flash EEPROM/E²PROM (Electrically EPROM) or FEEPROM/FE²PROM. Programming and erasing are done electrically. This is called *In-Situ* Programming (ISP). More generally, a component that can be programmed (via registers or internal memory) when it has been added to the board is referred to as In-System Configurable (ISC). The underlying board must of course have a suitable interface. Communication between the host and the target takes place via a traditional RS-232[1] asynchronous communication interface (*cf.* § 8.2.2 in Darche (2003)) using an integrated program or via a debugging interface (*cf.* § 2.2.5). The latter can be used for debugging. Figure 2.5 shows the ISP/PDI interfaces (Program and Debug Interface) used in the Atmel STK600 evaluation kit.

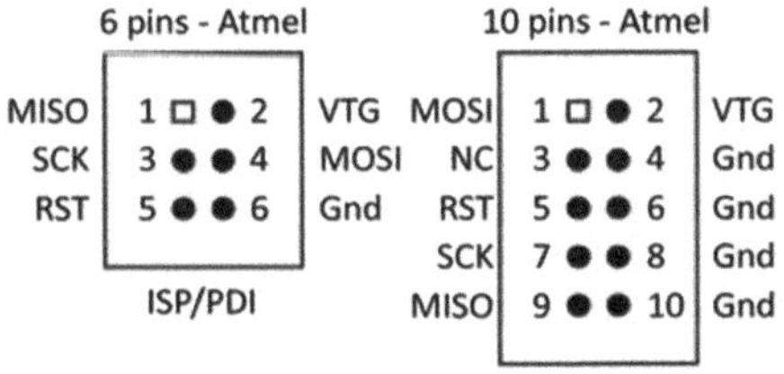

Figure 2.5. *ISP and PDI interfaces for the Atmel STK600 evaluation kit*

The IEEE 1532-2002 (IEEE 2003) standard specifies programming for configurable components. The state machine managing the programmable component has two primary states, system and test (Figure 2.6). The system state is used for normal operations (i.e. initial), even if non-intrusive commands can be executed. The test mode enables control of the component's I/O pins to allow programming of memory, for example.

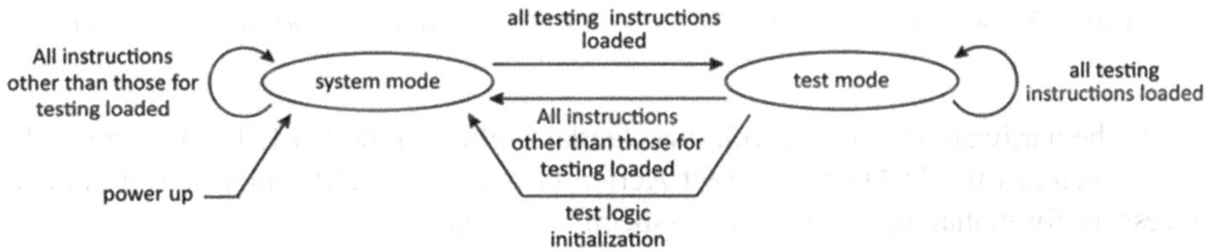

Figure 2.6. *Management state diagram compatible with the IEEE 1149.1 standard*

1 Controller associated with this interface: UART, for Universal Asynchronous Receiver Transmitter. Today, this interface is emulated via a USB interface.

The four standard-required instructions are ISC_ENABLE, ISC_DISABLE, ISC_PROGRAM and ISC_NOOP. The first two (de)activate the programming environment. The next enables system programming. The last makes it possible to put the system into idle mode (NO OPeration) in the case of parallel programming. The optional programming instructions are ISC_READ, ISC_ERASE, ISC_PROGRAM_USERCODE and ISC_DISCHARGE. For control and security, the optional instructions are ISC_PROGRAM_DONE, ISC_ERASE_DONE and ISC_PROGRAM_SECURITY. Instructions for accessing an address field or optional data are ISC_READ_INFO, ISC_DATA_SHIFT, ISC_SETUP, ISC_ADDRESS_SHIFT and ISC_INCREMENT. Ten standardized registers have been associated with these instructions. Figure 2.7 presents their relationships with the aforementioned registers.

	Target registers									
Instructions	ISC_Default	Device_ID	ISC_PData	ISC_Config	ISC_RData	ISC_Sector	ISC_Info	ISC_Data	ISC_Address	ISC_Inc
ISC_Enable	×			×						
ISC_DISABLE	×			×						
ISC_PROGRAM	×		×							
ISC_NOOP	×									
ISC_PROGRAM_USERCODE		×								
ISC_SETUP				×						
ISC_READ			×		×					
ISC_ERASE	×					×				
ISC_DISCHARGE	×					×				
ISC_PROGRAM_DONE	×		×							
ISC_ERASE_DONE	×		×							
ISC_PROGRAM_SECURITY	×		×			×				
ISC_READ_INFO							×			
ISC_DATA_SHIFT								×		
ISC_ADDRESS_SHIFT									×	
ISC_INCREMENT	×									×
	Required		Optional							

Figure 2.7. *Compatibility between standardized instructions and target registers*

At the hardware interface level, this standard relies on the IEEE 1149.1 standard, which includes the JTAG (Joint Test Action Group, *cf.* § 2.2.5) interface. It is also necessary for managing programming-specific signals.

2.2. Debugging

Debugging[2] for a processor-based system involves searching for and resolving faults and errors, both in hardware and software, during the development, tuning,

2 Real-time debugging is outside the scope of this work and will not be addressed.

production and running phases. It is important to distinguish pure hardware or software approaches from hybrid approaches.

2.2.1. *Evolution*

Hardware debugging and testing for microprocessor board(s) was initially carried out primitively using solid or blinking LEDs, or – a slight improvement – with messages on LEDs or LCDs (Liquid Crystal Display). A parallel port could be used to produce a binary number representing an error code, as was the case with the first IBM PC (Personal Computer, *cf.* § 3.2.1) during POST (Power-On Self-Test, *cf.* § 3.5.3). Some motherboard manufacturers still offer this functionality in the form of one or more seven-segment LED displays. These POST codes are sometimes associated with a set of audio beep codes. A final primitive approach is to program system memory, observe the application's operation and potentially debug it by iterating over these three steps as long as the system is not functioning correctly. This method is referred to as burn and learn, and is only applicable to small microcontrollers. Microprocessor system debugging has become complex with the integration of several levels of cache and acceleration mechanisms such as pipelines (this will be covered in a future book by the author on microprocessors). It is also important to note that debugging electronics that originally functioned at the board level are now integrated at the chip level. The complexity of debugging is still increasing.

At the software level, calls to message sending routines (internal system state) for signals or errors via the RS-232 serial connection (the traditional console) were inserted, either conditionally or unconditionally, in the application code, similar to the printf() debug print statement in C using preprocessor directives such as #if. Small programs can also generate observable events on an analog oscilloscope. For example, a sequence of `nop` (no operation, *cf.* § V4-2.8.5) instructions increments a sequence counter, which is observable on the address bus (*cf.* exercise V4-E2.6). Another example is a loop including a complementary state change[3] (toggle mode) for an I/O port pin, which will manifest via a rectangular signal. Professionals quickly realized the need for more powerful approaches.

2.2.2. *Functionality*

Debugging is relevant to every language level, from machine language to assembly and high-level source code (source level debugging). The debugger's

3 In terms of the Boolean algebra (i.e. alternating between 0 and 1).

functions can be divided into two groups, those for runtime control and those that enable system access. For the former, the fundamental functions are execution start and stop, single-step mode execution, and breakpoint placement, which can be synchronous with an instruction, simple (i.e. unconditional), complex or temporary (goto command), and their detection. Instruction tagging makes it possible to stop the execution of a program when the marked instruction is reached but not executed by the MPU and to return to debugging mode (*cf.* § 2.2.5). The breakpoint mechanism is based on a comparison of an address value or data depending on its type in relation to a reference. The breakpoint can be managed via hardware or software. In the case of variable-length instructions (the fixed-length case is trivial), it is useful for debugging to determine the instruction boundaries in the machine code (interruptible "at the instruction boundaries," *cf.* § V4-1.1).

For improved speed, as a complement to the traditional step-by-step execution mode, a special step-over mode makes it possible to execute subprograms in normal mode, while the rest of the program is running in step-mode. It is important to distinguish between three types of breakpoint: execution breakpoints, memory breakpoints (triggered by read/write or by a defined value) and I/O breakpoints on a port or event basis. Some debuggers also distinguish between the execution modes of user-mode debugging and kernel-mode debugging. The word synchronous means that there are no more instructions executed after the one immediately preceding the requested stop. With a complex breakpoint, it is possible to specify a range of addresses or to use a logical mask. A logical condition can also be used to trigger the stop. These functions for controlling execution then enable access to information. Thus, during an execution stop, they allow observation and modification of the microprocessor's internal (registers and internal memory) and external (memory areas) resources. It is also possible to place breakpoints on data. There are also watchpoints that can be set on an instruction, a value, an address reached in memory or the value of a register, which is identical to a breakpoint except that it does not stop execution but triggers the sending of a message sent to the debugging tool. It should be noted that the function at the root of a breakpoint or watchpoint is comparative. Monitoring instructions are especially important in the real-time domain where it is difficult to capture events. The difference between hardware and software breakpoints for instructions lies for the former in the fact that they are limited to RAM if the memory overlay functionality is not used. For an instruction breakpoint, also called an execution or program breakpoint, the debugger replaces the instruction from the target address with a rerouting instruction, generally a trap (*cf.* § V4-5.4), which stops the application and re-routes the instruction flow to the debugger. For the hardware version, specialized electronics detect the binary patterns that are stored in specialized registers in a wave, or, today, in the MPU. Both types are synchronous. Event and access breakpoints are asynchronous. Huges and Sawin III (1978) describe the technique for a software breakpoint.

Advanced capture functionality involves tracing the execution of instructions, access to data in memory and events such as hardware and software interrupts (*cf.* Chapter V4-5, i.e. exceptions) Tracing consists of real-time storage of a set of information (instructions, data, addresses, etc.) related to execution. It then allows a debugging tool to replay the execution, which renders the system externally observable. To increase execution speed, this information must be compressed on-the-fly. Its value lies in obtaining information prior to an error rather than just after, as is the case, for example, with a breakpoint, which then enables analysis at the error breakpoint. It requires significant memory resources, relative to the historical amount required. Execution traces are recorded by a logic analyzer or another suitable tool. To limit the message size (frame) and the transmission time, these can also be compressed.

Source-level debugging facilitates program tuning. The debugger can also grant access to the corresponding code in assembly language, which is useful when optimizing compiler-generated code. Furthermore, dynamic Performance Analysis (PA) can be provided for analysis; for example, the number of interrupt requests or calls to a subprogram, or the failure rate for the various cache levels. The debugger can also provide memory or I/O emulation.

Debugging becomes more complicated in the presence of an Operating System (OS) and, even more so, for real-time applications. The most powerful debuggers can trace system calls, for example.

Debugging a System on (a) Chip (SoC, *cf.* § V1-1.2 and V2-4.10) is the most complicated type. Debugging an embedded system is difficult because, unlike a computer, it does not usually have advanced peripherals such as a keyboard or monitor. Furthermore, the signals on the PCB are no longer available because the system has been integrated. To debug this kind of component, it is necessary to be able to control the execution (start, stop and step-by-step) and to manipulate (i.e. read/write) the MCU's[4] registers and memory from an external tool using a communication interface. To this functionality, we must add ROM emulation by RAM (loading) and real-time analysis of the instruction flow with tracing.

By building on the notion of class in the IEEE-ISTO 5001™-1999 standard (IEEE-ISTO 1999), four classes of debuggers can be defined. Class 1 provides basic functionality, that is, execution control (start/stop/step-by-step), with breakpoint and watchpoint insertion from which it is possible to access registers and, in memory, the variables and constants, and to read and write when necessary. The half-duplex

4 MCU for MicroController Unit or μC, *cf.* § V3-5.3.

dialog between the host and the target occurs via a test interface with a limited throughput (*cf.* § 2.2.5). The second class adds program and membership tracing (of a running function or process for an OS) via an auxiliary interface. This auxiliary port is shared with slow I/O, and as such has limited bandwidth. Class 3 enables read/write operations in memory and tracing of data writes in memory on-the-fly, that is, without stopping execution. I/O on the auxiliary port becomes fast. The last class authorizes memory substitution via the auxiliary port and watchpoint-triggered tracing. Data read tracing in memory is also possible. Table 2.1 reviews the incremental characteristics of these classes.

	Classes of debugger			
Characteristics	**1**	**2**	**3**	**4**
Runtime control	JTAG interface	JTAG interface	JTAG interface	JTAG interface
Communications	Cross-band (limited throughput)	Full-duplex (high throughput)	Full-duplex (high throughput)	Full-duplex (high throughput)
Auxiliary port	No	Yes shared with slow I/O pins	Yes shared with fast I/O pins	Yes shared with fast I/O pins
Trace storage	Not supported	Yes Execution via auxiliary port	Yes On-the-fly execution Data write tracing On-the-fly memory read/write (MEMR/MEMW) via auxiliary port	Yes On-the-fly execution Checkpoint triggers via auxiliary port
Data acquisition	Not supported	Not supported	Yes	Yes
Memory substitution	Not supported	Not supported	Not supported	Data read via auxiliary port Substitution trigger via watchpoint (optional)

Table 2.1. *Incremental characteristics of classes of debuggers*

2.2.3. Hardware emulators

A hardware emulator intervenes between the host and target system. It makes it possible to mimic the behavior of hardware. In the context of our study, a microprocessor is replaced by an external hardware device equipped with a specialized probe that will generate signals from the latter. It therefore emulates its internal operation and signals from the electrical and temporal points of view. By redirecting the microprocessor's functionality, it becomes possible to isolate it for better observation and control of its operation, which facilitates both tuning of the electronics board under development and its future maintenance. Its functions are mentioned above. It can also store a real-time execution trace. It can also provide a memory space in place of or complementary to the application's space (overlay memory). This memory area, which substitutes for the target system's memory, allows the user to reload the application code or to patch it (i.e. to apply a patch without having to repeat the software development chain, namely compilation, assembly and link editing) without having to program the microcontrollers or the system. This area can furthermore replace a memory area for a system that is not yet present. The exact name is "emulator on user circuit." Intel uses the term In-Circuit Emulator (ICE); Intel also developed the ICE-80, nicknamed the blue box. These emulators were initially connected to a system or a logic development station (per HP's terminology). For example, the ICE-80 was associated with the Microcomputer Development System (MDS) 80. Another well-known example is the HP 64000 workstation from Hewlett-Packard. This architecture is described in Kline *et al.* (1976), Krummel and Schult (1977), and Stiefel (1979). These systems were an expensive solution because the electronics were complex and because the systems were cumbersome. Today, they have been replaced by the microcomputer (Figure 2.8). An emulator can be referred to as universal, becoming specialized for a given microprocessor by adding a module containing the latter's electronics.

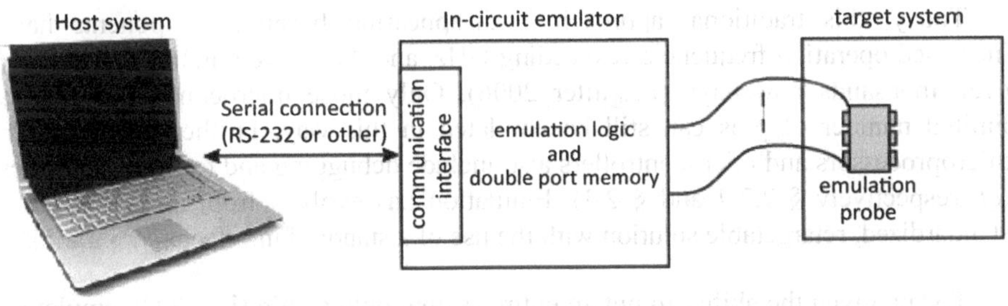

Figure 2.8. *Hardware emulation chain. For a color version of this figure, see www.iste.co.uk/darche/microprocessor5.zip*

Four operating modes can be distinguished: free-running mode, debugging, monitoring mode and step-by-step. In free-running mode, the target runs at full speed, and the debugger waits for a command to take control. In debugging mode, the microprocessor is stopped, and the debugger executes the debugging commands. In monitoring mode, the target runs at full speed, but the debugger monitors the breakpoints and watchpoints. As soon as one of these points is reached, the debugger switches to the preceding mode. In the last mode, the system alternates between monitoring and debugging modes for each instruction executed.

The core of the emulator is generally a microprocessor identical to the one on the board under development, with the exception that some of its internal control, state, and bus signals, etc., are externalized (bond-out chip). Internal registers such as the Program Counter (PC) are also accessible. All of these additional points of access make it easier to precisely know the component's internal state and to control its operation. They also enable implementation of function traces (*cf.* § 2.2.2). This type of component is provided by the manufacturers themselves to the manufacturers of emulators. The problem of commercial availability therefore arises, and its solution is a function of market size. Adding an emulator has an impact on the target's performance. For each signal, an additional delay is generally inserted, and an electrical charge is added (*cf.* § 3.6 in Darche (2004)). The emulator generally runs at a lower frequency than the real component and does not exceed hundreds of MHz, especially if the emulated microprocessor is replaced by equivalent discrete logic or, today, by FPGA (Field-Programmable Gate Array) type programmable logic circuits (*cf.* Chapter 4 and, more specifically, § 4.3.2 in Darche (2004)).

The logic (state) analyzer and the digital oscilloscope are devices for measuring and visualizing complementary signals, especially now that they include protocol analyzers.

Today, this traditional approach is disappearing because components have increased operating frequencies exceeding GHz, and they have hundreds of pins, or even thousands nowadays (i.e. after 2006). Only those microcontrollers with a limited number of pins can still be emulated in this way. Furthermore, modern microprocessors and microcontrollers also include debugging and testing electronics (*cf.* respectively § 2.2.4 and § 2.3). Emulation has evolved from an *ad hoc* to a standardized, retargetable solution with the use of a standard interface (*cf.* § 2.2.5).

Today, given the ability to put an entire system onto a chip (i.e. SoC), emulators are typically embedded. The ARM7/TDMI embedded processor was one of the first circuits with integrated ICE (EICE for Embedded ICE).

Finally, for reference, there are RAM and ROM emulators that have double access ports. The first is accessible by the target, and the second by a probe that can read, write or program depending on the memory type.

2.2.4. *Software debugging*

Historically, software debugging was done *post mortem* by analyzing memory contents (core dump) in the form of a paper printout (i.e. listing). The debugging monitor was the first embedded software (FirmWare or FW, *cf.* § 3.5.1), which assists with tuning. Today, it is important to distinguish between standalone software debuggers and those assisted by a hardware device that controls the processor.

2.2.4.1. *Debugging monitor*

Historically, microprocessor evaluation boards were equipped with software called a debug monitor for managing the board. This was a program resident in ROM (ROM monitor) for launching and tuning applications. It managed various controllers and peripherals for a microprocessor system (keyboard, displays, serial connection). The corresponding service routines were made available to the developer for their applications. It also made it possible to load, launch and update user programs. To do so, a command interpreter associated with a dedicated communication interface read command inputs and provided recording and branching for the corresponding system process. After execution, a report was displayed on the traditional output (displays, screen, etc.)

As examples, we will mention the JBUG/MINIBUG II and BUFFALO monitors (Bit User Fast Friendly Aid to Logical Operation) respectively for the Motorola MEK6800D2 and M68HC11EVB boards. MINIBUG is firmware that makes it possible to load and launch a memory dump. It can list the contents of registers and modify a memory location. Table 2.2 reviews the general characteristics of these early monitors. It should be noted that the monitor could be included in boot ROM. Thus, the notion of a monitor is found in microcomputers such as Macintosh or in SUN workstations, some of which included a basic monitor that provided terminal functions, allowing, among other things, reinitialization, MPU register and memory access, and the boot support definition (*cf.* § 3.5.3) with the help of input commands.

Characteristics	Generic debugging monitor
Software aspects	Firmware on target
Intrusive	Yes – monitor resident in user space User code patch for breakpoints Keyboard-display or access port required to communicate with the host
Code download	Yes
Execution commands (run/stop/step)	Yes
Code breakpoint	Simple only (generally software)
Data breakpoint	No
Access	Processor registers/variables
On-the-fly access	No
Observation point	No
Real-time tracing	No
Clock and timer	Can be used in step-by-step monitoring mode
Wait and stop modes	Monitoring mode not accessible in these two modes
Communications	RS-232 protocol, standard PC data rates
Cost	Low
Advantages	Simplicity/cost/broad family support
Disadvantages	Intrusive/only basic debugging functionality

Table 2.2. *Generic characteristics of a debugging monitor*

Table 2.3 shows the main characteristics for a representative monitor, the Motorola MCU HC08.

Characteristics	HC08 monitor mode
Input in mode	At least four generic I/O pins (V+ on IRQ signal)
Communications	RS-232 protocol (NRZ), standard PC data rates
Software	Firmware on target
Commands	Number = 5 Indirect access to MPU registers Access to memory only in monitoring mode
Breakpoint	One associated function
Clock or timer	Active in monitoring mode
Wait and stop modes	Monitoring mode not accessible in these two modes
Debugging connector	16-Pin MON08

Table 2.3. *Primary characteristics of the HC08's monitor mode (Motorola)*

It is important to distinguish between the monitoring functions of user mode and supervisor or monitor mode. The latter can be emulated if the microprocessor does not have execution privileges.

The monitor uses the board's resources, specifically the read-only (monitoring program) and random-access (workspace) memory spaces and some I/O controllers (serial, parallel, etc.). The monitor is generally installed with the interrupt vector table (*cf.* § V4-5.7) or, at least, the initialization vector (reset). Figure 2.9 shows an installation at the top of the address space for an MPU like the MC68xx with Motorola's BUFFALO monitor. The interrupt vector table can be found at the top of the address space (typical case for 8-bit MPUs) or at the very beginning.

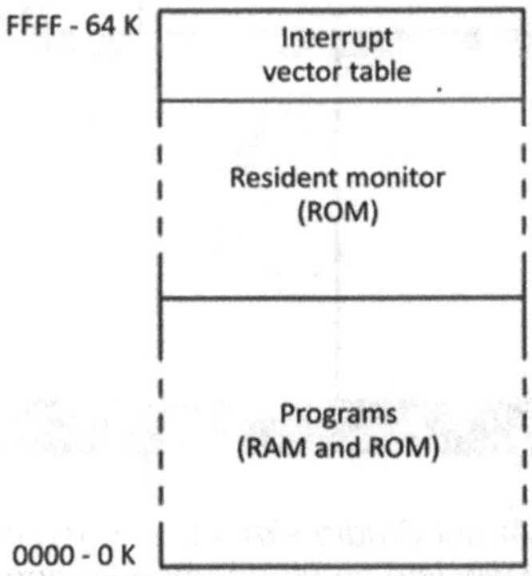

Figure 2.9. *Memory location of the monitor*

2.2.4.2. *Software and in situ simulators*

A simulator is a software that imitates a component's behavior. It makes it possible to test a user program without interacting with the hardware. The effectiveness of this approach is limited, but it enables debugging of the most serious errors. As with a debugger, the user can see on the display (Figure 2.10) the register set for the microprocessor, the memory areas including the stack, interrupt table, data area (i.e. variables) and the area for instructions. In the latter, the program exists in the source language, generally a High-Level (programming) Language (HLL) such as C, or in assembly. In the case of a microcontroller (Figure 2.10), the registers of the I/O controllers as well as the state of the peripherals are also shown. Besides than step-by-step and continuous modes, the program can also be executed in conditional mode (instruction or cycle counter, logical conditions) by simulating input–output controllers and memory such as the cache. It is also possible to set breakpoints. This can help detect invalid operation codes. Some simulators provide statistics such as the average number of loops or the number of times a variable is accessed; with this information, the user can optimize their program. This tool is provided in addition to others such as a hardware emulator (*cf.* § 2.2.3). It also offers obvious pedagogical utility.

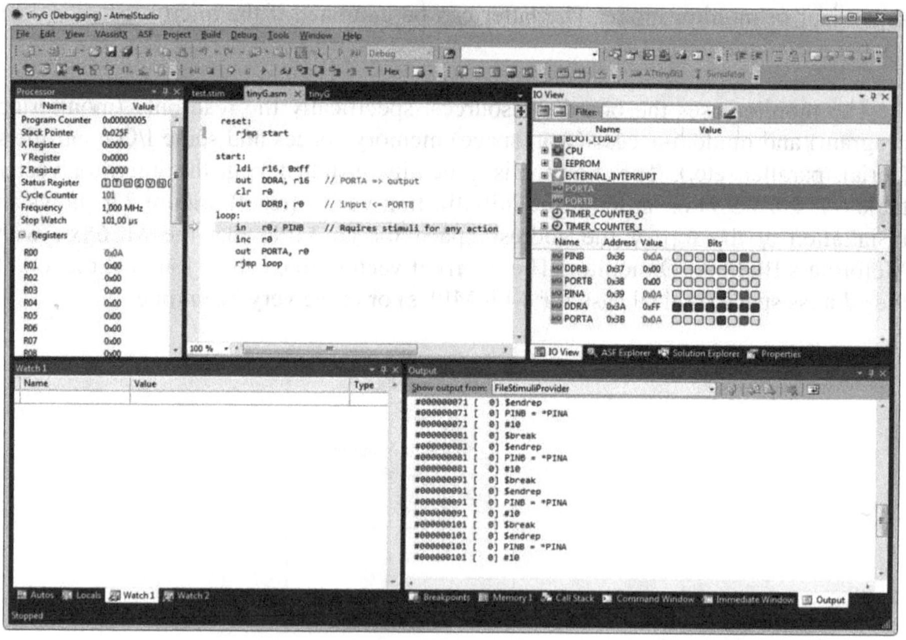

Figure 2.10. *The Atmel Studio simulator. For a color version of this figure, see www.iste.co.uk/darche/microprocessor5.zip*

One example of a mixed hardware–software simulator is the In-Circuit Simulator (ICS), which exists between a software simulator and a hardware emulator (*cf.* § 2.2.3). It is used for microcontroller-based development. The hardware part is made up of a traditional communication module, including the same microcontroller as the target, and is installed using a probe. It simulates operation by executing the application, but not in real-time. This allows an ICS to interact with the target's I/O, which adds new functionality. All software development tools and simulators can then be used.

Given the increasing complexity of hardware, and in order to facilitate hardware/software co-development, a high-level model can be constructed, for example, in C. The system state then exposes itself in the form of variables with unrestricted access, which facilitates testing. Thc primary disadvantage is that this approach does not take real-time issues into account.

2.2.4.3. *Software debugging*

Modern processors, beginning with the 16-bit generation of microprocessors (or 8-bit for microcontrollers), now include some of the functionality of a hardware emulator, such as a step-by-step mode. For example, the Intel x86 family includes a binary indicator called a Trap Flag (TF, *cf.* § V3-3.1.5.6) in its status register, which, when set to 1, puts the processor into step-by-step mode (trap into ICE). This mode traces a program's execution by requesting an interrupt (`into 1`) after the execution of each instruction. A software debugger can then take over to monitor execution. Previously, this functionality didn't exist; in the case of microcontrollers, an alternative was to use a clock that would trigger an interrupt request after a number c of cycles corresponding to the execution time of the next instruction (example from the BUFFALO monitor associated with the OC5 (Output Capture) clock in the MCU HC08). Moreover, software breakpoints can be inserted in the form of dedicated `int 3` interrupt requests, with the help of the two associated debug registers, DR6 and 7; in this case, the CCh machine code replaces the initial instruction. Other processors can use a traditional software interrupt request (e.g. HC11's `swi` in the Motorola BUFFALO monitor). For each request, there is a corresponding interrupt vector that contacts the start address for the debugger (interrupt handler). The disadvantage is that the user program is modified by replacing an instruction with a software interrupt request.

Today, this tool enables program tuning by helping find errors (bugs). It uses a testing interface (*cf.* § 2.2.5) that makes it possible to interact with the processor. Just as a system or a component, a microcontroller, for example, may integrate flash EEPROM; the debugger includes the functions of memory erasing, programming and checking. The debugger is sometimes called a "software emulator," but the term can be misleading because the additional integrated software circuits (hardware debugging support) are not the same as those in a hardware emulator. They only make it possible to access the circuit's I/O. It is better to refer to them as On-Circuit Debuggers (OCD)[5].

These provide the same functions as an emulator (*cf.* § 2.2.3). They make it possible to start and stop execution at any point using either conditional or unconditional breakpoints, as well as watchpoints. These points can be hardware- or software-related. For the latter, a comparison of the value of the sequence counter for the processor being analyzed with the stop address occurs in real-time. Their number is limited by the hardware. A software breakpoint is carried out by inserting a stop instruction (`break`, for example). It is therefore highly intrusive because it modifies the application code. The most advanced display source code in a high-level language as well as the corresponding code in assembly language (multilevel debugging), which requires a disassembler integrated with the object code. Figure 2.11 shows an example. As a final type of functionality, we should mention the patch assembler, which makes it possible to modify machine code during debugging without having to reassemble the source. A profiler makes it possible to generate static and dynamic statistics and performance measurements for application functions for the purpose of potential optimization, for example, but using assembly language in place of a high-level language.

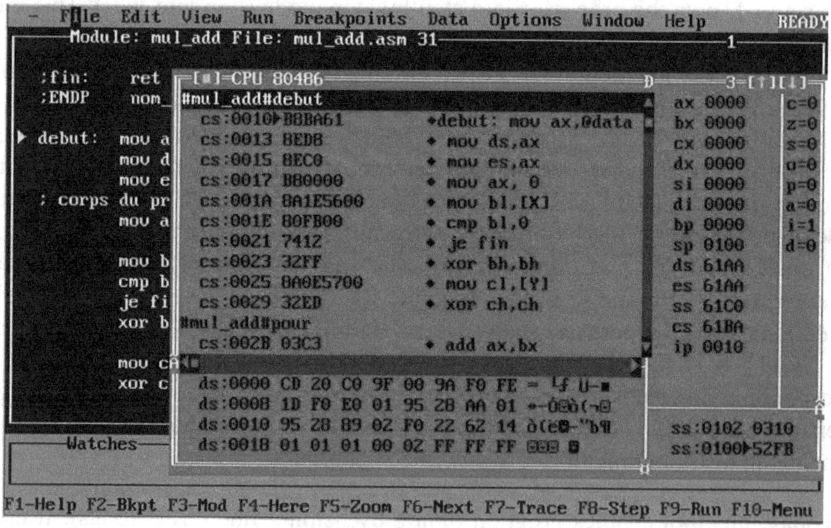

Figure 2.11. *The Borland TurboDebugger. For a color version of this figure, see www.iste.co.uk/darche/microprocessor5.zip*

5 The acronym ICD (In-Circuit Debugger) is rarely used. It is more specifically used in the commercial domain.

2.2.5. *Hardware support and debugging interfaces*

To facilitate debugging, an MPU can integrate specialized logic and registers. This is referred to as internal or on-chip debugging. In effect, mechanisms such as an internal cache mask the microprocessor's operation, because external observation of the bus is not correlated or, more precisely, is delayed from internal operations. Moreover, the technology of encapsulation has evolved, and the connection of a probe to capture the processor's signals has become difficult, while attaching a probe for an emulator with a BGA-type package (Ball Grid Array, *cf.* § 3.3.1 in Darche (2004)) is impossible.

An example is the x86 architecture beginning with the 80386, associated with the RF (Resume Flag), 8[6] Debug Registers (DR[7:0] used in part for hardware breakpoints in linear address space. DR[3:0] stores the addresses for four breakpoints simultaneously in memory space or in I/O (*cf.* § 2.3 in Darche (2003)). DR7 and DR6 are the registers for control and debug status respectively. RF suspends undesirable debugging exceptions while the debugger is being used. The OCD is transparent with respect to the MPU's operations. To implement this elegant solution, it is necessary to have an interface for communicating between the debugging tool and the system to be debugged. Solutions such as the traditional series, parallel and network interfaces are invasive because they use the target's resources. The transparent solution is to use a JTAG port designed for debugging.

Test interfaces make it possible to offer system access points, which are generally serial. They have debugging functions and, potentially, make it possible to program flash memory (in the case of a microcontroller, for example). A debugging module is inserted between the host and the target, just as with an emulator. It contains the debugging controller, which will generate the signals for the debugging interface. A set of debugging commands is associated with this interface. The host transmits commands and data, which are sent to the target via the debugging interface to be executed. If necessary, it receives the data and a state. The interface on the host side is traditional, either parallel, serial or network (Figure 2.12). We should mention the DSP56000's On-Chip Emulation (OnCE™, Motorola 1992), which enabled a specialized interface to interact with the DSP (*Digital Signal Processor*) in order, for example, to read from registers and memory. This DSP's debugging mode was therefore activated by a specific input signal called #DR, either on reset or normal execution mode. It should be noted that specific instructions – `stop` and `wait` – also enable this (*cf.* § V4-2.5.2).

6 The use of registers DR4 and DR5 is reserved by Intel.

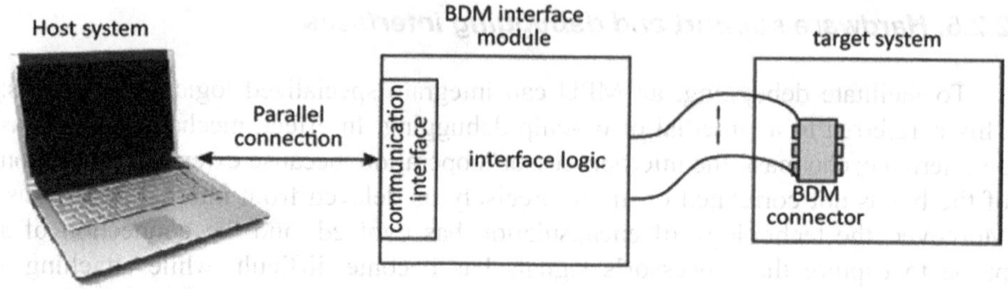

Figure 2.12. *Debugging link with BDM access (HCS08/RS08 target system). For a color version of this figure, see www.iste.co.uk/darche/microprocessor5.zip*

Each manufacturer initially specified its own interface (debug or probe port) with changes over the generations (families) of components. There are therefore proprietary interfaces such as Motorola's that use a full-duplex synchronous interface called BDM (Background Debug Mode). This enables communication with some of their microprocessors and microcontrollers, such as the M68HC08/12/16. It is made up of three signals (Figure 2.13), which are DSCLK (Data Serial CLocK), DSI (Data Serial Input) and DSO (Data Serial Output). Along with a set of commands, this interface enables read and write in a specific register or in main memory and checking of other functions connected to in-circuit debugging such as initialization, start at an address, insertion of a hardware breakpoint, or memory dump or load. The message exchange format is 17 bits. This interface was the first to be used for OCD.

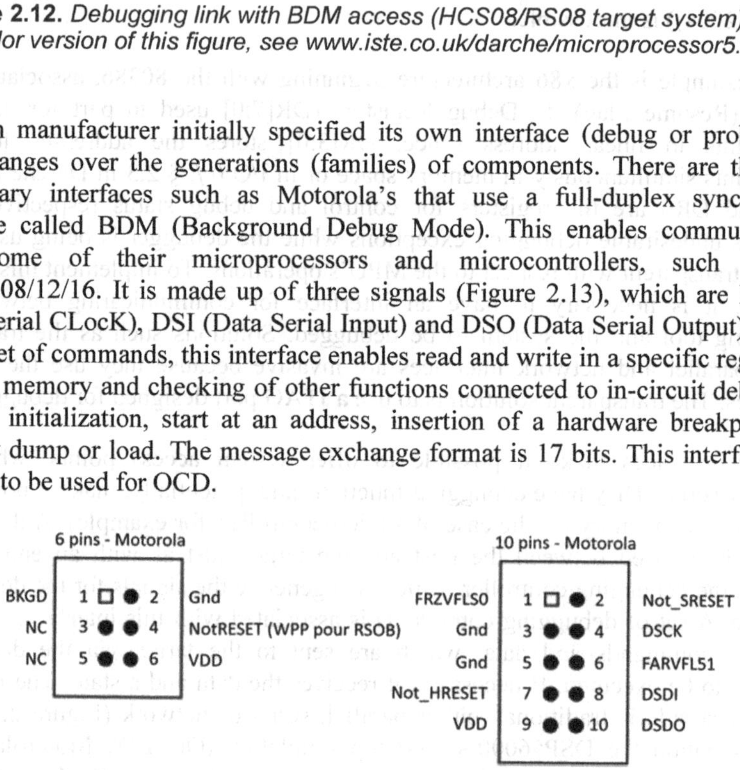

Figure 2.13. *BDM interface signals*

Table 2.4 shows the debugging mode characteristics associated with this interface. It should be compared with those from the monitor presented in § 2.2.4.1 (Table 2.3).

	Background debug mode (BDM)	
Characteristics	**HCS08**	**RS08**
Input mode	Single dedicated pin (BKGD) required (no V+)	Single dedicated pin (BKGD) required (no V+)
Serial communication	Dedicated protocol	Dedicated protocol
Software aspects	No firmware on target	No firmware on target
Hardware aspects	External debugging module BDC (Background Debug Controller) registers Outside user address space	Interface module BDC registers Outside user address space
Commands	Number = 30 17 active in BM and 13 non-intrusive Direct access to MPU internal registers in active BM Non-intrusive access to memory in BM and during execution of a user program	Number = 21 10 active in BM and 10 non-intrusive One in both modes Direct access to MPU internal registers in active BM Non-intrusive access to memory in BM and during execution of a user program
Breakpoint	One hardware and one instruction trace Two breakpoints supported by the external module	One hardware and one instruction trace
Clock and timer	Inactive in active BM	Inactive in active BM
Wait and stop modes	Authorized background access	Authorized background access
(Debugging) connector	6-Pin BDM	6-Pin BDM

Table 2.4. *Characteristics of the background debugging mode for Motorola HCS08 and RS08 microcontrollers*

We should mention, in no particular order, the AMDebug™ (12 pin, improved JTAG) interface, the Arm Debug Access Port (DAP) interface, Atmel's aWire, PDI, and debugWire, NEC's N-Wire, Motorola's OnCE™, SDI (Serial Debug Interface), and SPI (Serial Peripheral Interface), and finally, TI's MPSD (Modular Port Scan Device).

The IEEE 1149.1 (IEEE 2013) standard specifies the test logic to be included in an Integrated Circuit (IC) to enable testing of the interconnections between ICs once they have been soldered to the circuit board or to the component itself. It also enables observation and modification of the component's activity in normal operation. This testing architecture is non-invasive. For this purpose, a general-use, serial-type Test Access Port (TAP), also referred to as a JTAG interface (Joint Test Action Group), in reference to its origins, is also specified to enable communications between the board being observed and the test system (Figure 2.14). Internally, additional circuit-limiting logic enables the application of test vectors and the receipt of results. Its architecture is sufficiently generic to make it possible to leverage its initial test functionality for debugging. It is possible to monitor and alter the processor's internal logic state. Thus, the set currently offers three functions, which are boundary scan testing, *in situ* debugging (OCD) and memory programming.

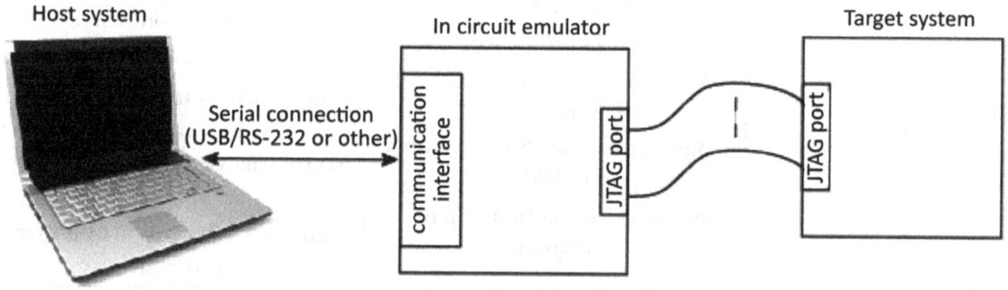

Figure 2.14. *Debugging connection using a JTAG port. For a color version of this figure, see www.iste.co.uk/darche/microprocessor5.zip*

The principle of peripheral boundary scan testing relies on a chain of serially linked test modules (Figure 2.15). It enables testing of the continuity of the external connections between logic systems and between a system's logic circuits, as well as also internally at the logic block level. A test module is made up of two kinds of registers, test data registers and instruction registers. Instruction registers receive the commands to be executed. Management is granted to a controller (TAP controller), which manages the testing phases based on a 16-state state machine (*cf.* § 3.7.3 in Darche (2002)). There are four, or optionally five, interface signals. There are input and output instructions and test data signals respectively TDI (Test Data Input) and TDO (Test Data Output), the test clock signal TCK (Test ClocK), the TMS (Test Mode Select) mode signal and, potentially, the nTRST (Test ReSeT) initialization signal, which is active at a low level. TDI makes it possible to enter instructions and data serially into the circuit, sampled at the rising edge of TCK. TDO makes it possible to extract data. TMS makes it possible to change the state machine's state. TCK is the sampling signal for the TMS signal and the strobe signal for data and instructions. The transfer speed is not greater than 100 MHz, which limits real-time debugging. The asynchronous nTRST signal, when it is at a low level, enables initialization of the controller.

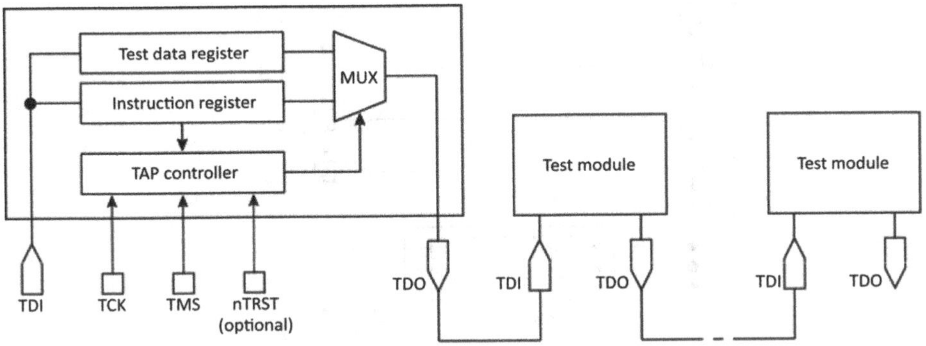

Figure 2.15. *A chain of test modules.*

Internally, communication relies on a chain of Boundary-Scan Cells (BSCs) surrounding the circuit being tested, corresponding to one cell per I/O pin on the original circuit (Figure 2.16). The scan path principle is a widespread testing technique. It was proposed by Williams and Angell (1973) for reading and setting a state in a logic system. The testing technique relies on a derivation of signals with the aid of multiplexers and flip-flops, the latter of which can be placed in series to form a Shift Register (SR, *cf.* § 3.5.3 in Darche (2002)) for the transport of test information. This peripheral chain forms the Boundary-Scan Register (BSR), which is one of the five testing data registers, which is required, as are the IDR (IDentifier Register) and the Bypass Register (BR).

The IDR register provides an identifier for the manufacturer, the circuit and its version, in 32-bit format. As the name suggests, the BR register enables direct passing of a bit entered by TDI to the TDO. The two others, which are optional, are the scan-path and BIST registers (*cf.* § 2.3). As an example, consider the BSR for the Motorola MPC8xx family of microcontrollers, which has a 475-bit format for testing all of its signals[7] (McEwan 2002).

7 Except for three analog signals, XTAL, EXTAL and XFC.

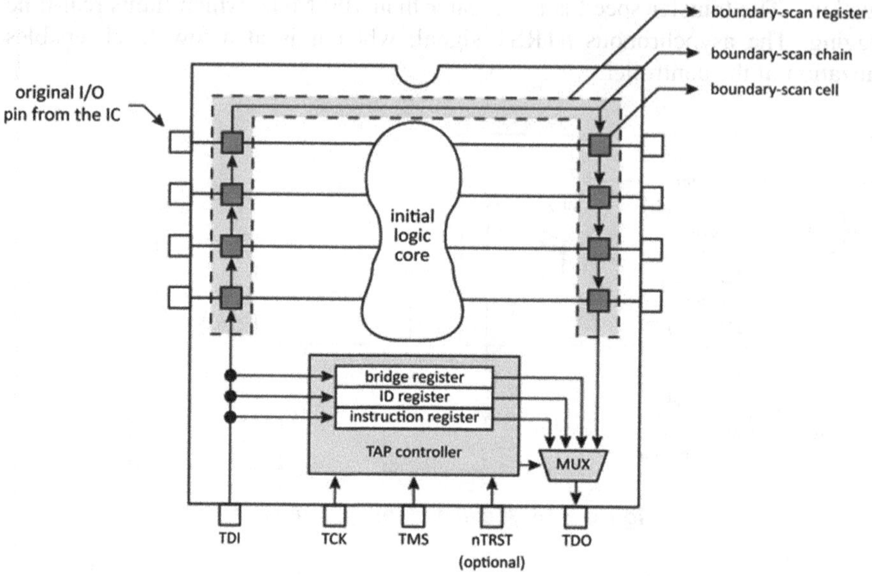

Figure 2.16. *JTAG logic around the initial core*

A boundary-scan register cell makes it possible to observe and control the DUT's (Device Under Test) external input and output signals. Using a multiplexer controlled by the TAP controller (Figure 2.17(a)), the incoming or outgoing signal on the pin of the circuit to be tested can pass through transparently or can be taken charge of by the flip-flop network. Figure 2.17(b) shows an example implementation. The M2 multiplexer selects the signal path, whether normal or under test. The M1 multiplexer selects the input signal for observation or the one to be inserted. It is then taken over by two serial flip-flops, R1 and R2. The first (R1) is the capture flip-flop and the second (R2) is the update flip-flop, which is optional. Each cell is linked by Scan_In/Scan_Out signals, thus forming the boundary-scan test path.

There are 18 instructions that are either public or private. There are four required public instructions, which are BYPASS, SAMPLE, PRELOAD and EXTEST. The optional public instructions are INTEST, RUNBIST, CLAMP, IDCODE, USERCODE, ECIDCODE, HIGHZ, INIT_SETUP, INIT_SETUP_CLAMP, INIT_RUN, CLAMP_HOLD, CLAMP_RELEASE, TMP_STATUS and IC_RESET. Each manufacturer can also define its own instructions. A standardized language called BSDL (Boundary-Scan Description Language) has been proposed.

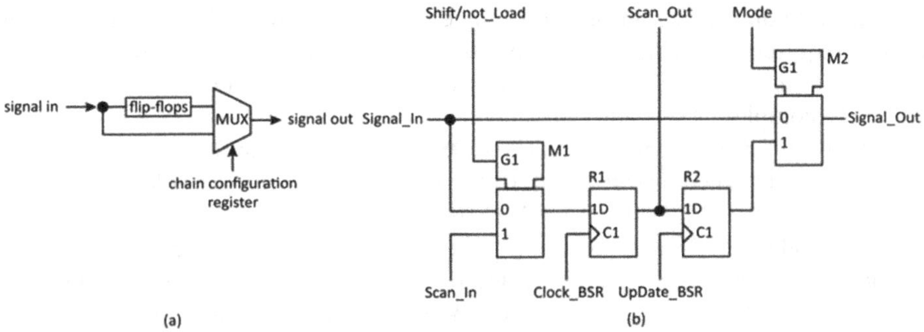

Figure 2.17. *A typical boundary-scan register cell (a) with an example implementation (b)*

Initiated by Hewlett-Packard in 1989 (Parker 2002), it is based on the standardized VHDL (Very High Speed Integrated Circuit, VHSIC) Hardware Description Language (IEEE 2000, also *cf.* Darche (2002)). More details on its internal operation are given in Maunder and Tulloss (1990).

Each manufacturer proposes a connector that includes at minimum the standard's four or five signals. In practice, the connector has 10–20 pins (Figure 2.18). The SWD (Serial Wire Debug) interface from Arm® is a subset of the JTAG interface. It should be noted that the RTCK (Return TCK) signal is not standardized. We should also point out the ITP (In-Target Probe) interface from Intel.

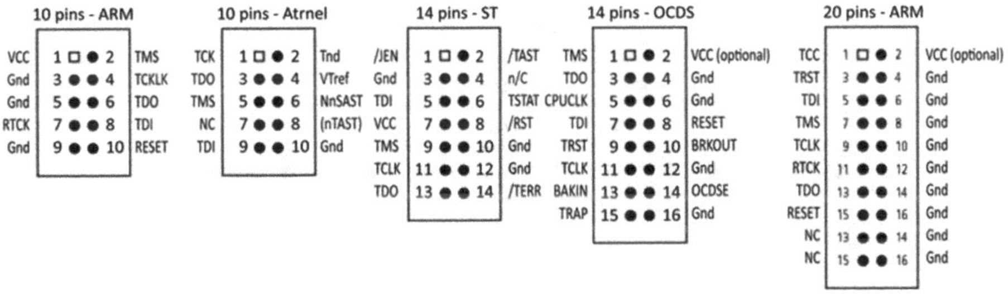

Figure 2.18. *JTAG connector examples*

With the spread of SoCs, which make it possible to integrate an entire system onto a chip, testing and debugging have become excessively complicated. Previously, all signals between chips were accessible at the board level and could be observed by traditional measurement instruments such as logic analyzers or oscilloscopes, or by the aforementioned debugging systems. Today, these tools must adapt to technological progress and become integrated at the chip level. Debugging an SoC requires observing, in addition to the operation of one or more processors, the internal bus (*cf.* § V2-4.10) and the system's peripheral circuits, which makes it extremely complex. The traditional approach using the JTAG interface is no longer suitable. An On-Chip Debug System (OCDS) embedded in the SoC is required. This approach is presented in Stollon (2011). To respond to this problem, the IEEE 1500™ standard (IEEE 2005; IEEE/IEC 2007) revisited the idea of the IEEE 1149.1 standard to adapt it to SoCs. Around the component, we find peripheral logic, shown in Figure 2.19. A three-signal interface enables communication. These latter are referred to as Wrapper Serial Output (WSO), Wrapper Serial Input (WSI) and Wrapper Serial Control (WSC). The Wrapper Boundary Registers (WBR) enable access to the cores (i.e. logic blocks) whose operations are controlled by events (four options: delay, capture, update and transfer). Instructions such as `bypass`, `preload`, etc. are passed via the Wrapper Instruction Register (WIR). They determine the state of the wrapper. DaSilva *et al.* (2003) describe their operation. WBY (Wrapper BYpass Register) enables a short-circuit of the test logic.

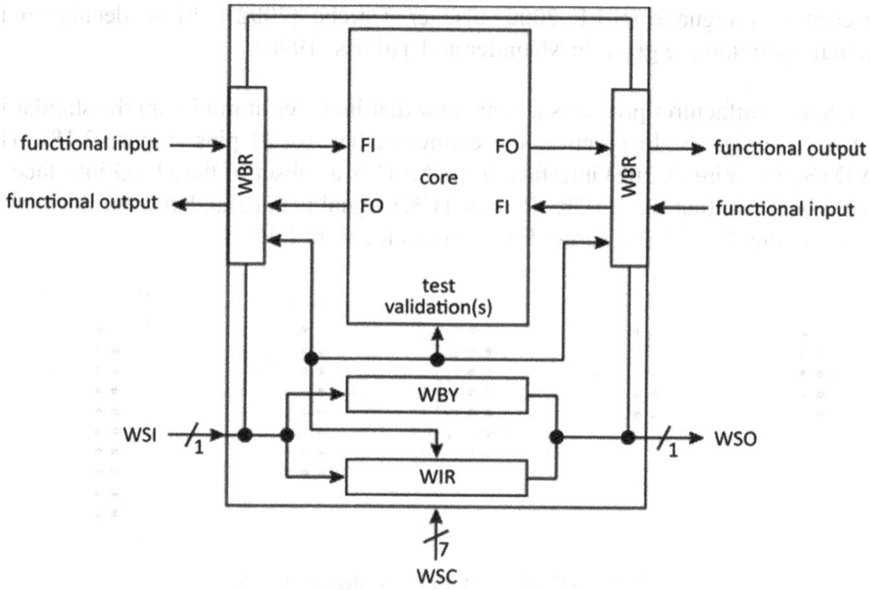

Figure 2.19. *Components required by the standard*

To capture a trace, that is, to capture information on-the-fly without stopping the system, the aforementioned interfaces are not suitable. An additional trace port interface must be added, as proposed in the Nexus standard (IEEE-ISTO 1999), which named it the auxiliary port (unidirectional). Another solution is to use internal memory. This port is a proprietary solution such as the Debug Communication Channel (DCC) or the trace port for ETM (Embedded Trace Macrocell) embedded logic from Arm®.

Other proposals exist, such as the MIPI (Mobile Industry Processor Interface) Alliance and the OCP-IP (Open Core Protocol International Partnership). They are not described here because they are outside this work's scope. Vermeulen *et al.* (2008) take stock of these standardization activities.

In the XScale® architecture, Intel mixes JTAG technology with the monitor-type approach by downloading an image of a debugging exception manager into a reduced-size cache (2 KiB) reserved for this purpose. The JTAG interface is here diverted from its testing role to serve only as the serial communication interface. The manager is called using the traditional debugging interrupt. We have here the typical functionality of a debugger with read/write of registers and memory, management of hardware instruction and data breakpoints, as well as software-type breakpoints for instructions and management of the instruction trace buffer.

The OCD's operation is transparent with respect to the MPU. No resources external to the MPU are used. This makes it possible to test hardware, initialize the system and debug applications under development. An additional optional feature is programming or "flashing[8]" the main programmable memory with a data check, which is useful for production testing. Moreover, all resources accessible by the MPU are accessible by the OCD. Thus, it is possible to leverage its functions for production, for example, for circuit calibration.

2.2.6. *Remote debugging and virtualization*

Today, debugging is done remotely over the network. Therefore, the circuit to be debugged can be located on another network or on the same one as the host. The gdb tool (GNU[9] Project debugger) enables a crossover debugger on the host while the software agent (gdbserver) is located on the target (Figure 2.20). The concept of a virtual machine can be applied to both development and debugging. In conjunction with the QEMU hardware emulator, the gdb tool makes it possible to debug in the Linux system space.

8 Term derived from the Flash EEPROM memory technology.

9 GNU for GNU's Not UNIX.

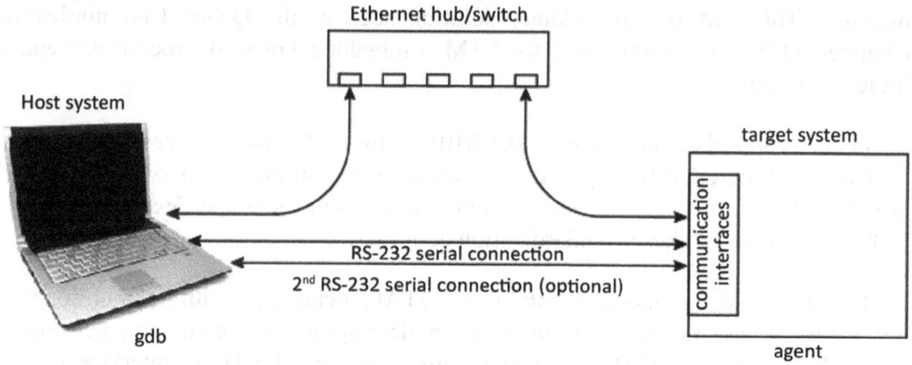

Figure 2.20. *Redirected debugging with gdb. For a color version of this figure, see www.iste.co.uk/darche/microprocessor5.zip*

ARMulator is an ISA (Instruction Set Architecture) simulator for the ARM® family. It also enables communication with the target so it can be observed. One example of free software is SkyEye.

2.2.7. *Summary*

A development system is made up of a set of software, hardware and hybrid tools. These include the tools in the development chain (assembler, compiler, link editor[10], loader), software and hardware debuggers, the emulator and the logic analyzer.

Table 2.5 shows early examples of debugging. *Postmortem* debugging consists of stopping the MPU and dumping its main memory (RAM) to analyze its contents. The iterative development method for firmware, called burn and learn programming, consists of writing new firmware, writing it to system ROM in test mode and observing the system as it runs. From this observation, particularly during a crash, the developer can analyze the errors to understand and correct problems.

10 That is, a linker.

	Debugging techniques			
Characteristics	***Postmortem***	**Burn and learn**	**Debug monitor (ROM)**	**Software simulation**
Real-time execution	Yes	Yes	No	No
Software aspects	Memory dump	-	Firmware	Specialized software
Hardware aspects	None	None	Possible keyboard/displays	None
I/O management	No	No	Yes, but limited	No, to be limited
Intrusive	No	No	Yes runs in user space communication interface of the target in use	No
Code download	No	No	Yes	Yes
Execution commands (run/stop/step)	No	No	Yes	Yes
Code breakpoint	No	No	Simple only (generally patched software)	Unlimited, complex
Data breakpoint	No	No	No	Yes
Complex breakpoint	No	No	No	Yes
Access to processor registers/variables	No	No	Yes	Yes
On-the-fly access	No	No	No	Yes

	Debugging techniques			
Characteristics	***Postmortem***	**Burn and learn**	**Debug monitor (ROM)**	**Software simulation**
Watchpoint	No	No	No	Yes
Tracing	No	No	No	Yes
Timer	No	No	Can be used in step-by-step monitoring mode	No
Wait and stop modes	No	No	Monitoring mode not accessible in these two modes	No
Connection (i.e. cabling)	No	No	Simple	No
Communications	No	No	RS-232 protocol standard PC data rates	No
Loss of processor pin	No	No	No	No
Cost	Low	Low	Low	Low
Advantages	Simplicity/cost	Simplicity/cost	Simplicity/cost/ widespread family support	Simplicity/cost
Disadvantages	Very limited functionality	Very limited functionality	Only basic debugging functionality Requires a target with resources (memory, communications)	Real-time aspects are impossible

Table 2.5. *Comparison of functionality between various debugging tools (1/2)*

Table 2.6 shows more advanced tools. These include in-circuit simulation, in-circuit emulation and in-circuit debugging.

	Debugging techniques		
Characteristics	**In-circuit simulation (ICS)**	**In-circuit emulation (ICE)**	**OCD**
Real-time execution	No	Yes	Yes
Software aspects	Specialized software	Specialized software	Specialized software
Hardware aspects	Processor identical to the target	External module with probe	External module
I/O management	Yes but limited	Yes	Yes
Intrusive	No	No	No
Code download	Yes	Yes	Yes
Execution commands (run/stop/step)	Yes	Yes	Yes
Code breakpoint	Complex (hardware type), unlimited	Complex (hardware type), unlimited	Optional, complex (hardware type), limited
Data breakpoint	Yes	Yes	Yes, optional
Complex breakpoint	Yes	Yes	Yes
Access to processor registers/variables	Yes	Yes	Yes
On-the-fly access	Yes	Yes, optional	Yes, optional
Watchpoint	Yes	Yes	Yes, optional
Tracing	No, to be limited	Yes – real-time	Yes – real-time, optional
Timer	No	No	No

Characteristics	Debugging techniques		
	In-circuit simulation (ICS)	**In-circuit emulation (ICE)**	**OCD**
Wait and stop modes	No	Yes	No
Connection (i.e. cabling)	Complex	Complex	Simple
Communications	Serial protocol	Dedicated protocol	Dedicated serial protocol
Loss of processor pin	No	No	Yes
Cost	Low	High	Low, medium if tracing
Advantages	Simplicity/cost	Covers all debugging needs	Excellent quality/price ratio
Disadvantages	Specific to a family of components Real-time aspects are impossible	High cost Specific hardware to a component Operating frequency limited	Dedicated internal logic Additional interface at the component I/O level

Table 2.6. *Comparison of functionality between various debugging tools (2/2)*

Chen *et al.* (2002) categorize modern in-circuit debugging (*circa* 2000) into two functional modes: ForeGround Debug Mode (F(G)DM) and BackGround Debug Mode (B(G)DM. In the former, the debugger controls the processor's operation as would an external emulator. In the latter, the microprocessor executes the application until a known event such as a breakpoint is reached, for example. The debugger then transitions to foreground execution; after debugging operations and on command, it returns to the background. Moreover, there are three approaches to debugging: hardware, software and hybrid. The authors have summarized these approaches in Table 2.7. There are completely software-based solutions (approach no. 1, *cf.* § 2.2.4) and completely hardware-based solutions (approach no. 2 *cf.* § 2.2.3), as well as two hybrid solutions. Kao *et al.* (2008) revisited the topic to modify debugging for the SoC environment. For this situation, there are three kinds of architecture: centralized, distributed and hierarchical (Huang and Kao 2000).

	Approaches	Software emulation (FSBS)	Hardware emulation (FHBH)	Hybrid emulations	
	No.	1	2	3	4
Modes of operation	**Foreground (FDM)**	Software	Hardware	Software	Hardware
	Background (BDM)	Software	Hardware	Hardware	Software
	Advantages	Flexible, easy to modify	Complex breakpoint real-time execution	Complex breakpoint real-time execution (= FHBH) Flexibility of FDM implementation	Flexibility Easy to modify (= FSBS) Software size for FDM smaller
	Disadvantages	System memory space occupied firmware on target (monitor) Longer time to detect breakpoint and to return to user mode	Large number of management logic ports dedicated, unmodifiable solution, inflexible and expensive	Slower FDM execution	Hardware solution is more expensive than FSBS
	Examples	Buffalo monitor (Motorola HC11) monitor mode (Motorola HC08) instruction breakpoint and step-by-step mode for x86	Hardware emulator (ICE) ARM7TDMI	x86 IA-32/64 architecture debugging registers (Intel)	BDM interface (Motorola) core with BSC cells and JTG interface associated with software interrupt instruction

Table 2.7. *Classification of debugging approaches (Chen et al. 2002). F: Foreground, B: Background, S: Software, H: Hardware*

2.3. Testing

A microprocessor is, at the very least, a complex VLSI (Very Large-Scale Integration)-type integrated circuit and is therefore difficult to test. This challenge must be overcome when life is on the line, which is the case in the military or space sectors. Aside from those conducted during design, tests were at first done at each manufacturing step, then during the component's burn-in or stress tests. To facilitate this process, or even to make it possible, the concept of Design for Testability (DfT) has been employed. Testing covers embedded memory (RAM and ROM), including internal data and instruction caches, and other much more limited memory areas such as registers (*cf.* § V3-3.1), First In, First Out buffers (FIFO, this will be covered in a future books by the author on memories and storage), etc. In § 2.2.5, we presented the IEEE 1149.1 standard, an access method belonging to the DfT approach and based on a test path (access or scan) that enables access to the logic circuit's internal nodes, with testing carried out using an external device. Manufacturers of advanced microprocessors now integrate internal self-test logic (BIST for Built-In Self-Test, Agrawal et al. (1993a, 1993b) and *cf.* § 3.9 in Darche (2004), which makes it possible to determine if the circuit is functioning correctly. For more information on this broad subject, see Wang *et al.* (2006) and Navabi (2010).

Errors during execution of an instruction can be caused by misuse, a component design bug or an unreferenced instruction. An example of the first case is division by zero. For the second, consider the `fdiv` floating-point division bug in the Intel Pentium® 5 discovered in 1994. An unreferenced instruction is an instruction code that is not documented in the official documentation. In the best case, it has no effect (equivalent to a `nop`). In the worst case, it creates a temporal reliability problem in programs that use it. For example, there were hidden instructions in MOS Technology's MCSD6502 MPU.

2.4. Conclusion

We have described the software development chain for a microprocessor system application. Software and hardware debugging tools were shown. Early debugging tools and practices such as *postmortem* analysis and the iterative burn and learn approach are no longer suitable today. Increasingly complex and fast MPUs require integration of testing and debugging logic as close to the core as possible. Internal debugging (OCD) and associated communication interfaces such as the JTAG interface were then examined. To conclude, we presented an overview of assembly language. This language enables symbolic manipulation of instructions, data (variables and constants) and addresses. High-Level (programming) Languages (HLL) have taken their place in embedded systems thanks to progress in storage and compilers. But it is still used in specific circumstances such as to optimize memory use and execution time, for system or hardware interfacing, or for teaching computer architecture. To learn more about this language for a specific processor, Streib (2011) presents an example from the Intel world.

The next chapter shows the microprocessor in the earliest hardware environments that were widely distributed, namely microcomputers.

PART 2

Part 2 describes the digital system engendered by the invention of the microprocessor; namely, the microcomputer. Two exemplars have been chosen to represent this category – the Apple II and the IBM Personal Computer (PC). The latter's successive motherboard architectures are described in detail. Two components are studied in depth: one hardware component – the chipset, and one type of firmware – the BIOS (Basic Input/Output System).

3

Changes in the Organization of the Earliest Microcomputers

This final chapter, which provides a partial conclusion to this volume, shows the evolution of the organization of the first microcomputers. From the earliest days of the microprocessor in 1971, there were evaluation boards such as the ones described in the previous chapter. Then, the microcomputer appeared, and the market for them exploded with the second generation (i.e. family version). Since the release of the Personal Computer (PC) in 1981 by IBM (International Business Machines Corporation), which refocused the PC for the business environment, motherboard architecture has not stopped evolving and taking advantage of the progress of integration in the domain of microelectronics. In particular, we saw the rise of the concept of the chipset. First, we will describe a second-generation microcomputer, the Apple II. Then, we will examine the evolution of the PC. The notion of chipset is elaborated. The architecture of modern motherboards is described. This chapter concludes with a discussion of low-level software from the firmware (FW) perspective.

NOTE.– The goal of this chapter is not to exhaustively describe the different architectures as would a specialized text about a particular model. It is to show the progression of architectures, even when focusing on a subsystem or a key feature.

3.1. Apple II

It is not possible to talk about the microcomputer without describing the Apple II as a worthy exemplar or as a support for the 2.0 or 2.5 generation of microprocessors (*cf.* § V3-4.3) or MPUs (MicroProcessor Units). It represents the first family microcomputers. Its motherboard (Figure 3.1) is a classic implementation from the era, based on TTL (Transistor–Transistor Logic, *cf.* § 2.3.2

in Darche (2004)) discrete logic circuits and LSI (Large-Scale Integration) circuits, including the MCS6502 MPU.

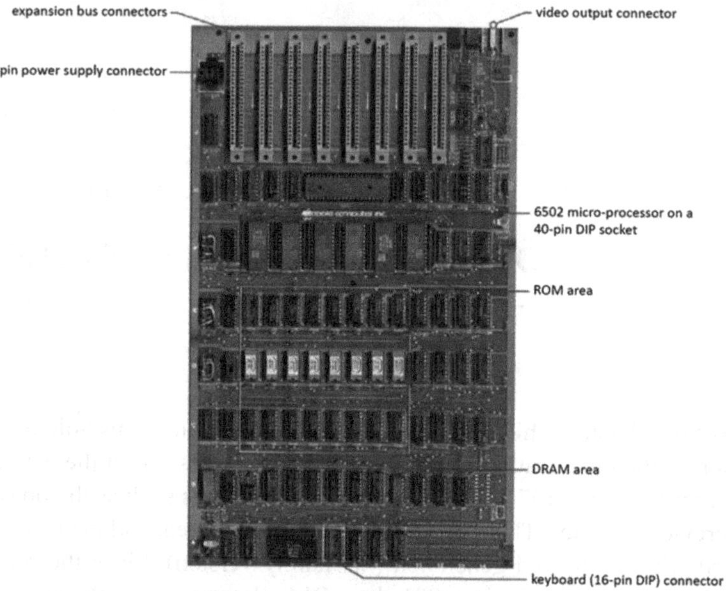

Figure 3.1. *Original Apple II motherboard (Rev. 0). For a color version of this figure, see www.iste.co.uk/darche/microprocessor5.zip*

The Printed Circuit Board (PCB) was made of (iconic green) FR4-type epoxy resin (Flame Retardant). The components used THT (Through-Hole Technology), which is analogous to the current SMC (Surface Mounted Component, *cf.* § 3.4.2) technology. With it appeared the extension connector "slot" designed to accept a daughter board, also referred to as an expansion slot, which sat perpendicular to the motherboard (eight available slots). Abundant discrete logic ensured the management (refresh, *cf.* § 5.2.2 in Darche (2012)) of DRAM (Dynamic Random Access Memory, shown in red in Figure 3.1), with a 48 KiB capacity (memory component reference 4116), and of the video display. The eight ROM (Read-Only Memory) components, each of which had a capacity of 2 KiB, are above the DRAM area. They contain a monitor (*cf.* § 2.2.4), a mini-assembler or disassembler, and a BASIC (Beginner's All-purpose Symbolic Instruction Code) interpreter for whole numbers called Integer BASIC[1] by its developer, Steve Wozniak (nicknamed Woz). The other integrated circuits are contained in DIP (DIL (Dual-In-Line) Package) 14- and 16-pin packages, providing glue logic (*cf.* § V3-2.3).

1 It was originally assembled by hand!

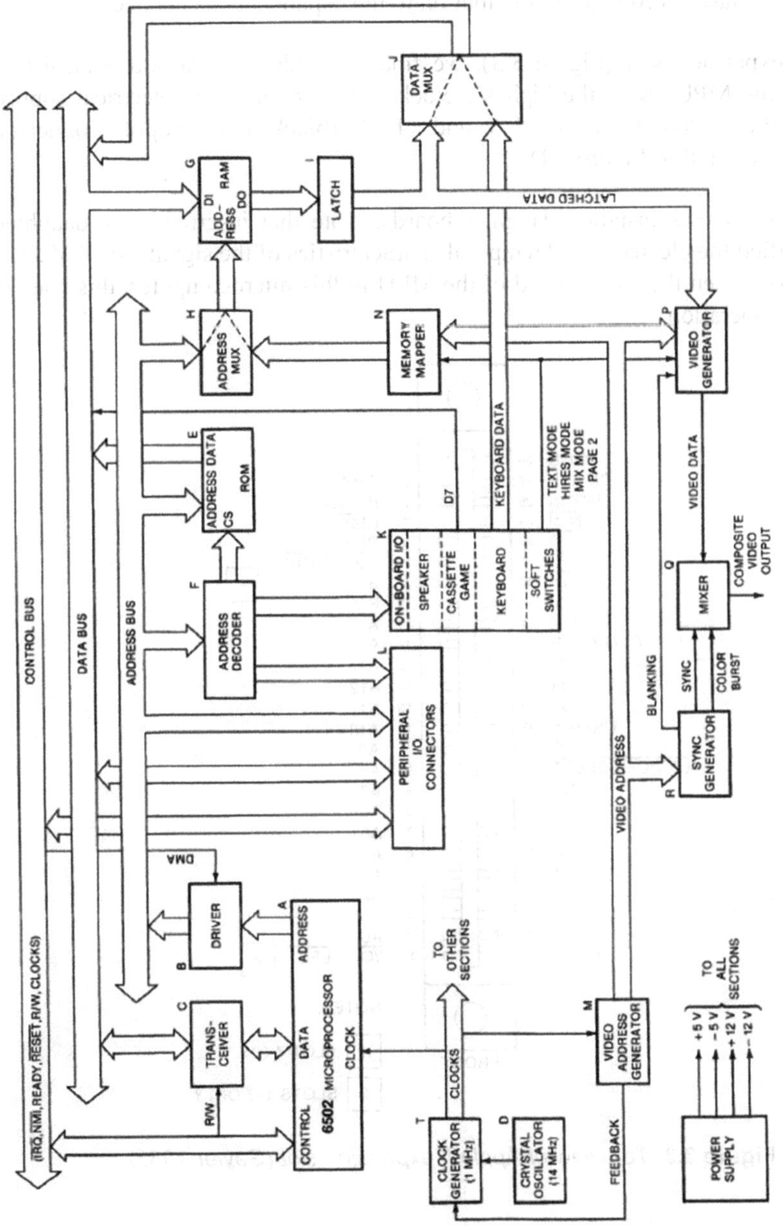

Figure 3.2. *Apple II block diagram (Gayler 1983)*

Figure 3.2 shows the general organization of this computer. The address and data bus signals (*cf.* Chapter 3) are simply amplified using 8T28 electronic buffers (*cf.* § 3.4.1 in Darche (2004)) for distribution to the expansion bus connectors.

In the expansion slot (Figure 3.3), we find the address, data and control bus signals for the MPU. Note the biphasic clock signals ϕ_0/ϕ_1. The electrical supply values for this generation were ±5 V and ±12 V (black power supply connector shown at the top left in Figure 3.1).

Figure 3.4 shows installed daughter boards. Note that insertion of a daughter board modified the electrical and temporal characteristics of the signals (*cf.* § V2-3.3 and V2-3.6). Given the clock speed of the MPU in this microcomputer, this had no effect on its operation.

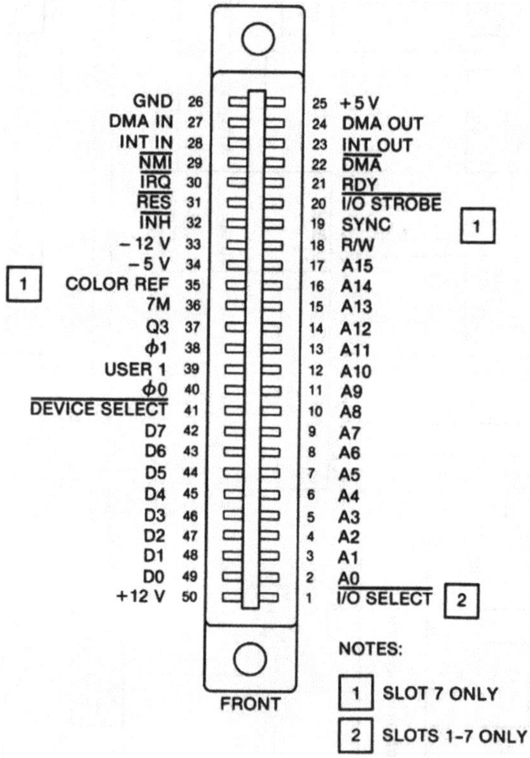

Figure 3.3. *Top view of Apple II expansion slot (Gayler 1983)*

Figure 3.4. *Installed daughter boards (unknown source). For a color version of this figure, see www.iste.co.uk/darche/microprocessor5.zip*

An innovative feature was that all of the system's clock signals, that is, the two signals from the 6502, DRAM refresh signals and video signals were generated from a master clock (f = 14.3 MHz). As a result, all of the subsystems that used them were synchronized. Dynamic memory was periodically swept to generate the video signal, which also refreshed it. This function constituted half of 1 µs MPU cycle, with the other half reserved for program execution. An oscillator built on an integrated NE555 timer circuit was responsible for modifying the video signal to make the cursor blink!

With regard to Input/Output (I/O), there was an interface for the keyboard, the speaker, the game port (joysticks and push buttons), and the audio tape recorder that provided mass storage (*cf.* § 7.2 in Darche (2003)). To read the voltage at the potentiometer terminals, the NE555 timer, generally used in monostable or astable mode, was adapted to create an Analog-to-Digital Converter (ADC). The software management routing was made responsible for this function. For more technical information, see Gayler (1983).

3.2. IBM PCs

The IBM professional line is the origin of the microcomputer, as we know it today. Three of their models made their mark on the IT industry. These were the 5150 (original PC), 5160 (XT) and 5170 (AT) models. We will now take a look at them.

3.2.1. *The original PC*

The Personal Computer (PC) was introduced by IBM in 1981. The history of this project at IBM is provided in Bradley (2011), Goth (2011), Bride (2011) and Singh (2011). Its open architecture is the foundation for modern microcomputers, which are referred to as “PC compatible”. It used the Intel 8088 MPU, which had a 16-bit internal architecture, an 8-bit external interface and a clock speed of 4.77 MHz. As with the Apple II, the clock speed was derived from a master oscillator running at 14.31818 MHz (divided by 3). The video signal (color burst) was obtained by dividing the same reference by 4.

It had a main memory capacity that ranged between 64 and 640 KiB (16 Ki x 1-bit DRAM reference 4116) and 40 KiB of ROM. The (EP)ROM ((Erasable Programmable) ROM) contained the BIOS (Basic Input/Output System), which included the Power-On Self-Test (POST, *cf.* § 3.5.3), the BASIC interpreter, the I/O routines, the 128-dot video pattern (i.e. ROM characters) and a bootstrap loader. The base RAM component was 16 Kib (type 4116). Each memory word had a format n of 9 bits, with the additional bit reserved for error detection via parity bit checking (EDC for Error-Detecting Circuit/Code). The controller alerted the user by generating an NMI (Non-Maskable Interrupt, *cf.* § V4-5.2) that enabled the display of a blocking message on the display. Figure 3.5 shows the motherboard.

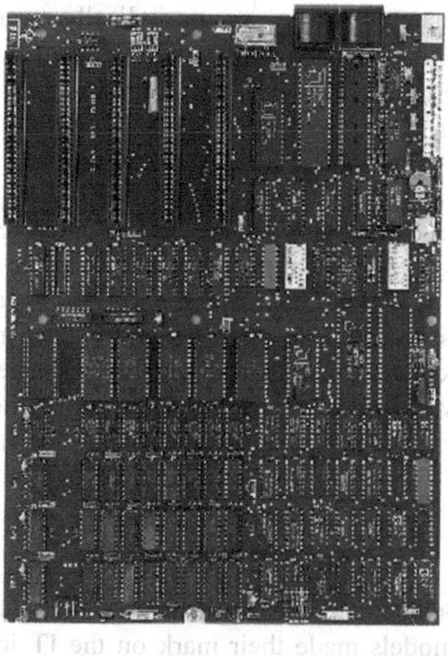

Figure 3.5. *IBM 5150 (original PC) motherboard. For a color version of this figure, see www.iste.co.uk/darche/microprocessor5.zip*

Figure 3.6 describes the position of the components on the motherboard. The (V)LSI components include DMA controllers (DMAC for Direct Memory Access Controller, Intel 8237A), Programmable Interrupt Controllers (PIC, Intel 8259A), timers (Intel 8253), parallel I/O controllers and Programmable Peripheral Interface (PPI, Intel 8255A) controllers. Glue logic belonging to the TTL family surrounds the memory and these controllers.

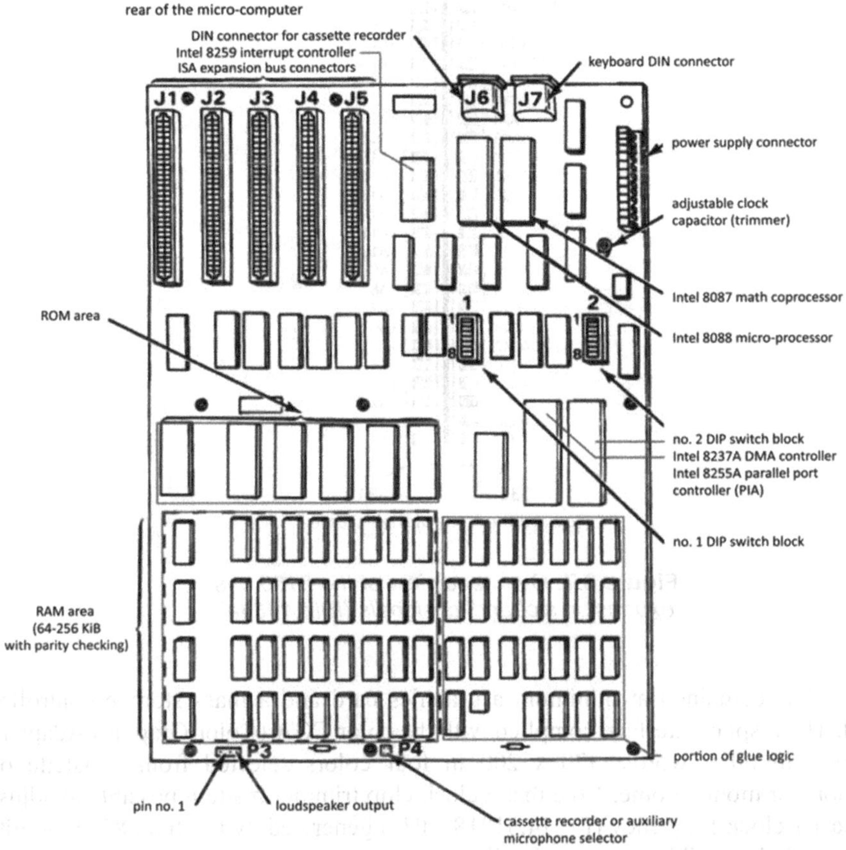

Figure 3.6. *Overview of the IBM 5150 motherboard (IBM 1984a). For a color version of this figure, see www.iste.co.uk/darche/microprocessor5.zip*

Five ISA (Industry Standard Architecture, *cf.* § V2-4.5) bus expansion slots made it possible to configure the machine for the user's needs by selecting expansion cards. As in the Apple II, the signals on this bus are buffered signals from the MPU (*cf.* § 3.4.1 in Darche (2004)).

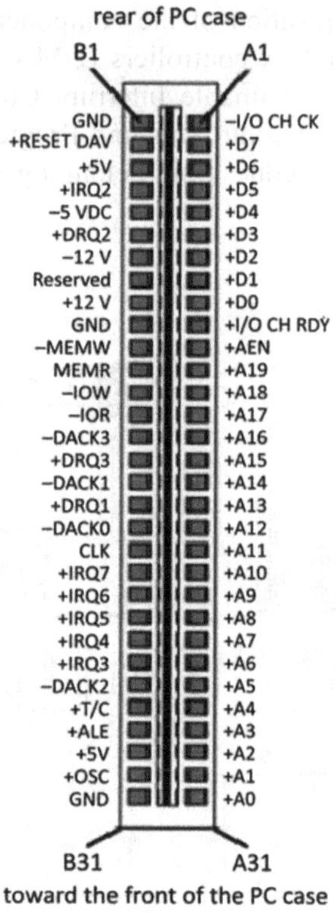

Figure 3.7. *Overhead view of the 5150 bus expansion slot and its signals (IBM 1984a)*

The bus contained at minimum a graphics card and a mass storage controller board. The display interface complied with the color CGA (Color/Graphics Adapter) standard, in other words, 320 x 200 in four colors selected from a palette of 16 colors, or monochrome. Note that a clock chip trimmer made it possible to adjust the master clock frequency (f = 14.31818 MHz) generated by the Intel 8284 circuit, which made it possible to match the display colors (at the time, a Cathode Ray Tube (CRT)) more precisely to the NTSC (National Television Standards Committee, f = 3.579545 MHz) color burst signal sent to a composite monitor. It contained two Floppy Disk Drives (FD for Floppy Disk and FDD for FD Drive, *cf.* § 7.2.2 in Darche (2003)) with a 5 1/4" format and a 360[2] KiB capacity (double density

2 Originally 160 KiB.

format), from which the MS-DOS (MicroSoft Disk Operating System), which became the foundation for the first version of Windows, could be loaded. Note the presence of two circular DIN (*Deutsches Institut für Normung*) five-pin connectors (IEC 60130-9), which provided ports for an 88-key keyboard and a peripheral device, an external audio cassette recorder providing mass storage. This was a holdover from the first two generations of microcomputers and was subsequently abandoned.

3.2.2. *The XT*

The next model, which was released in 1983 (the IBM 5160), also called the PC/XT or just the XT (eXtended Technology), was an improved PC with the same MPU running at the same clock speed. Among the improvements were a Hard Disk Drive (HDD) with 10 MiB capacity in 5 1/4" format, which used an ST-506 parallel interface from Seagate. The number of expansion slots was increased from 5 to 8, and the RAM increased to 640 KiB, although each memory word kept the 9-bit format, for the same reasons as its predecessor. Figure 3.8 shows the motherboard. Note that a version containing a 286 microprocessor running at a clock speed of 6 MHz, with the reference number of 5162 (XT/286), was released in 1986.

Figure 3.8. *IBM 5160 motherboard (PC XT). For a color version of this figure, see www.iste.co.uk/darche/microprocessor5.zip*

Figure 3.9 describes the position of the components on the motherboard. We can see there almost all of the same components that were in the original PC. However, the audio tape recorder interface was removed. The expansion slot signals were identical to the 5150, with the exception of the B8 pin, which carried the -CARD SLCTD (card selected) signal activated by a card (open collector type output, *cf.* § 2.3.2 in Darche (2004)) inserted into the J8 connector to indicate that it was selected (read only).

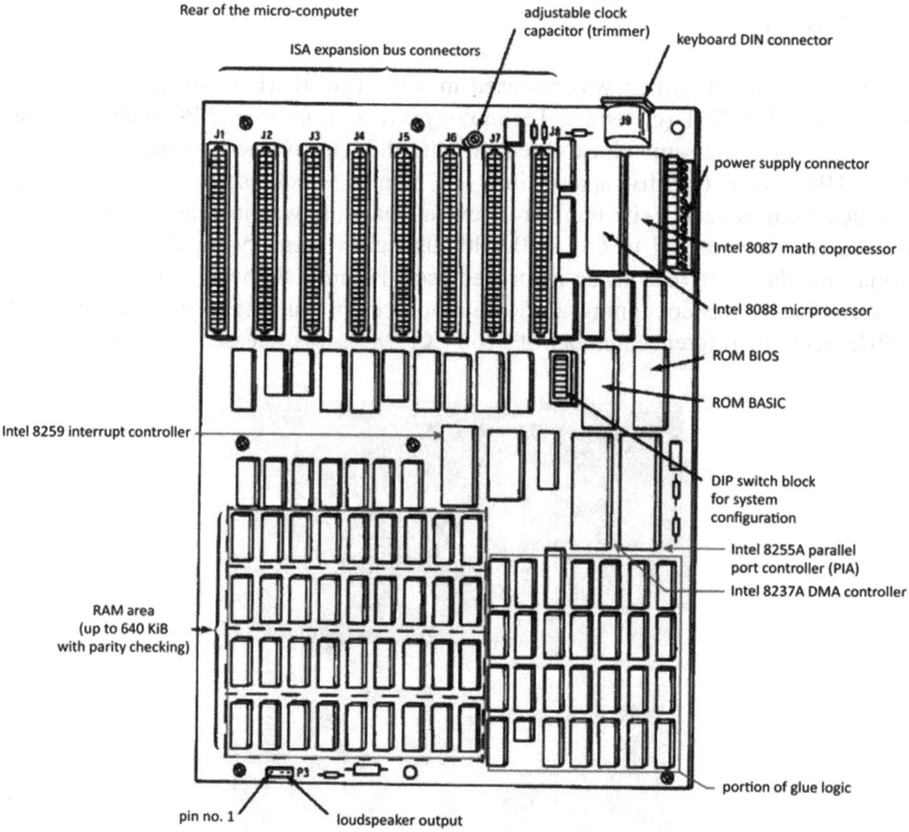

Figure 3.9. *Overview of the PC XT motherboard (IBM 1983). For a color version of this figure, see www.iste.co.uk/darche/microprocessor5.zip*

Note that, in this version of the machine, the POST (*cf.* § 3.5.3) is responsible for determining the RAM capacity. Previously, this was determined by reading a range of eight DIP switches (reference SW2).

3.2.3. *The AT*

One year after the XT, the IBM model 5170, also called the PC/AT or simply the AT (Advanced Technology), was architected around an 80286 microprocessor, a true 16-bit microprocessor clocked at 6 MHz, which made it possible to extend the address space to 16 MiB. Two addressing limitations existed with the preceding models. First, the limitation was physical, since the Intel 8088 microprocessor had an address bus size m of 20 bits (i.e. 20 address lines), which limited the address space to 1 MiB (= 2^{20}). The area between 640 KiB and 1 MiB was called the Upper Memory Area (UMA). Thanks to a memory management driver, it was possible to allocate Upper Memory Blocks (UMBs). The native Microsoft DOS (Disk Operating System) was therefore designed only to manage this conventional memory area. Additional memory belonging to this MiB could be used as a "RAM DISK", in other words, as a mass storage to speed up access and to increase throughput. To enable a larger space, two approaches were implemented: expanded and extended memory.

The first, called augmented or expanded memory, used the bank or page switching approach[3]. The idea was to add a RAM area above this megabyte that was divided into pages of 16 KiB (=64 KiB), each with 16 KiB. The size varied from 4 MiB to 32 MiB depending on the model. Only a four-page frame was accessible between 832 KiB and 896 KiB. Note that this approach had already been implemented in the Apple II with 16 KiB pages (Saturn 128 KiB memory board). Four concurrent technologies were on the market: LIM, EMS, EEMS and XMA. The first was developed conjointly by Lotus Software, Intel and Microsoft, which based its acronym on the companies' initials (EMS 3.0 – 1985). The Expanded Memory Specification (EMS) 3.2 was released in September of the same year and offered a memory space of up to 8 MiB (4 pages of 16 KiB). A software manager called EMM.sys (Expanded Memory Manager) needed to be added to the OS in the form of a driver. Subsequently, a version of the specification called EEMS (Enhanced Expanded Memory Specification), proposed by Ashton-Tate Research, Inc. (AST; for the initials of founders Albert Wong, Safi Qureshey and Thomas Yuen) and Quadram, extended the window to 64 pages. The latter was incorporated

3 Note that the term "page" does not refer to an implementation of the Virtual Memory (VM) pagination mechanism (this will be covered in a future book by the author on memories).

into LIM EMS 4.0 (August 1987), which offered up to 32 MiB of expanded memory. In 1986, IBM proposed its own specification using the acronym XMA (eXpanded Memory Adapter) for its expansion board, which offered 2 MiB of memory.

With the release of MPUs with greater physical address space, such as the 16 MiB (=2^{24} bytes[4]) available in the Intel 80286 and 386SX and the 4 GiB (= 2^{32} bytes) in the 80386, the previous technique became obsolete. Memory above the first MiB was still called "extended memory" for the purpose of distinguishing this first 1 MiB of space from the rest of the memory. The associated technology, called XMS (eXtended Memory Specification), was proposed by the LIM Association and AST Research. Version 2.0 was published in July 1988, and 3.0 was released in January 1991. The first 65520 (= 2^{16} - 16) bytes of extended memory accessible in real addressing mode[4] formed the High Memory Area (HMA). This range is the result of the difference in the management of the address space ranging from FFFF:0010 to FFFF:FFFF in the 80286 MPU compared to the 8088/86, since the latter could not go beyond a MiB. Figure 3.10 maps the PC memory space (concept explained in § 2.6.3 in Darche (2012)).

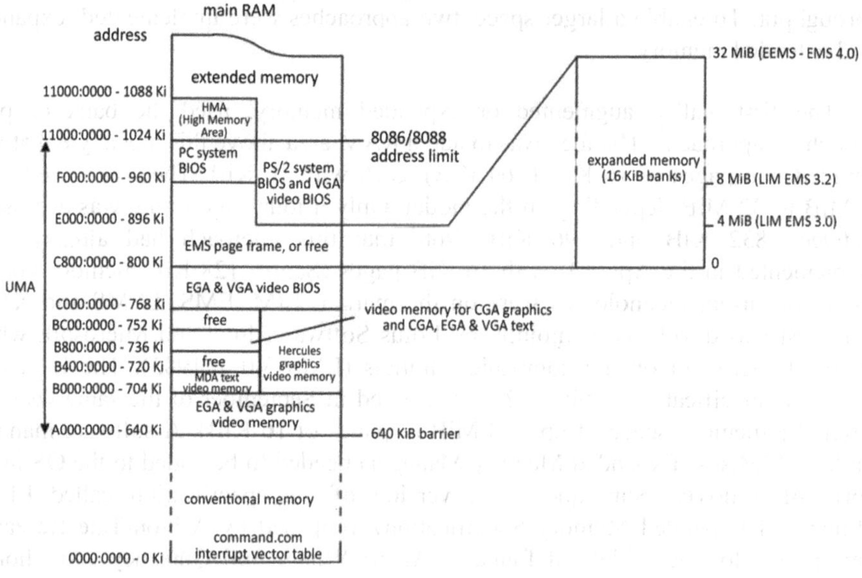

Figure 3.10. *Topography of main memory (Figure 2.110 in Darche (2012) completed)*

4 This mode was accessible only after initialization of the 80286.

Figure 3.11 shows an AT type 1 motherboard.

Figure 3.11. *IBM 5170 type 1 motherboard (PC AT). For a color version of this figure, see www.iste.co.uk/darche/microprocessor5.zip*

Figure 3.12 shows the position of the components on the printed circuit. Main RAM was composed of four banks of 9 DRAM, reference 41128, with a unit capacity of 128 Kib.

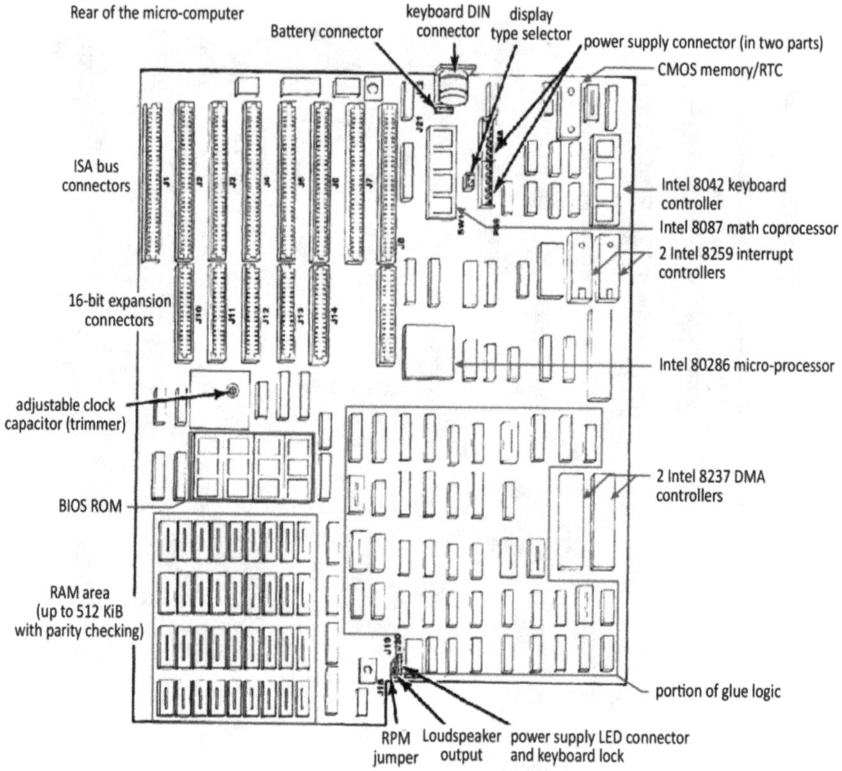

Figure 3.12. *Overview of the PC AT type 1 motherboard (IBM 1984b). For a color version of this figure, see www.iste.co.uk/darche/microprocessor5.zip*

The type 2 motherboard was narrower (Figure 3.13) thanks, particularly, to the reduction of the DRAM to two banks of 41256 integrated circuits (256 Kib).

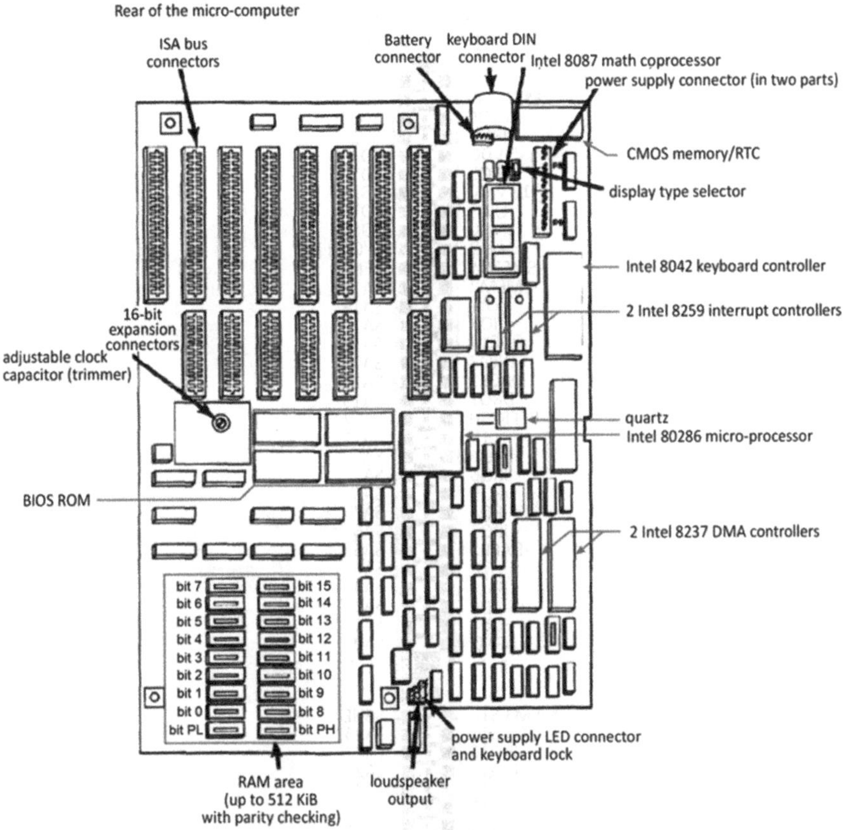

Figure 3.13. *Overview of the PC AT type 2 motherboard (IBM 1986). For a color version of this figure, see www.iste.co.uk/darche/microprocessor5.zip*

The ISA (Industry Standard Architecture) bus expansion format was extended to 16 bits using the name "AT bus", thanks to the addition of a second connector (a shorter extension from the first connector), which thus enabled upward compatibility with the ISA bus. Figure 3.14 shows its signals.

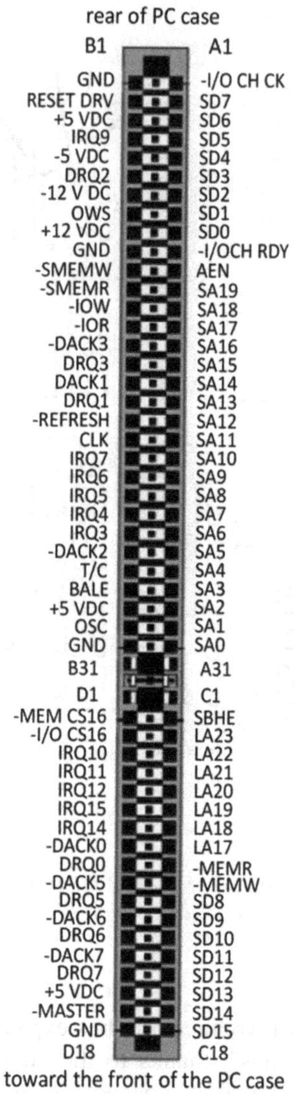

Figure 3.14. *ISA bus expansion slot extended to 16 bits and its 16-bit expansion signals (seen from above)*

Figure 3.15 shows the organization of this microcomputer. Electronic buffers amplify the MPU's signals on the bus to distribute them to the expansion slots. They can also be bidirectional.

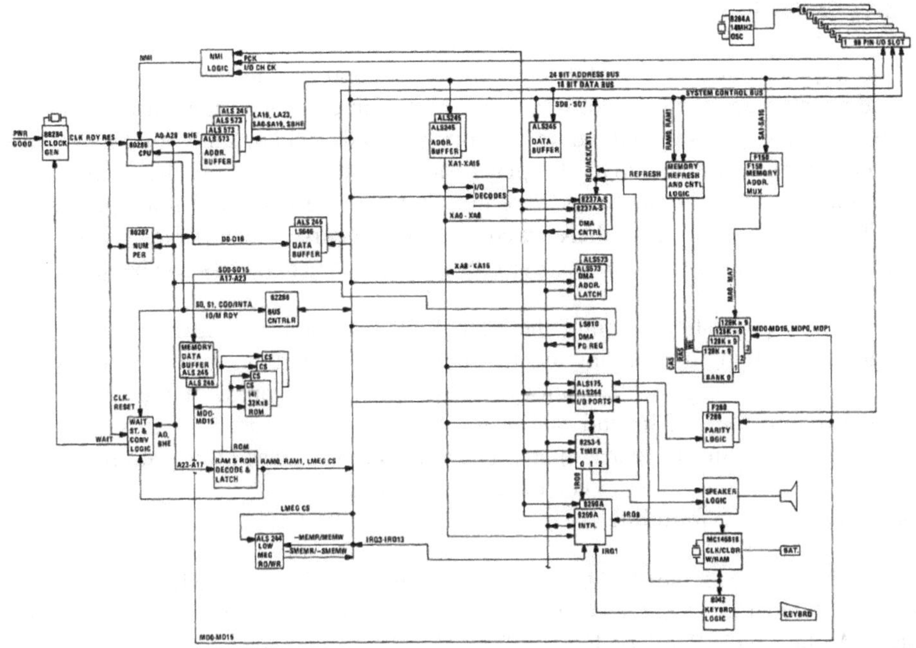

Figure 3.15. *AT block diagram (IBM 1984b)*

The capacity of the Floppy Disk (FD), which retained its flexible, square cartridge (i.e. enclosure), increased to 1.2 MiB (high density) in the same format as the XT. The hard disk drive unit, still in 5 1/4" format, increased to a 20 MiB capacity. Processing power increased by a factor of 4 compared to the XT (0.8 compared to 0.2 MIPS, *cf.* § V4-3.4 for the unit of performance measurement).

3.3. Chipset

With progress in integration, glue logic around the MPU, memory subsystems and I/O controllers were integrated into VLSI (Very LSI) chips, with the whole being called a chipset. After defining this component, the PC architectures that use it will be described.

NOTE.– This study runs until around the year 2000, and will be completed in subsequent works.

3.3.1. *Definition*

A chip set integrates digital logic circuits around the microprocessor (address decoding, etc.) and the I/O controllers into a set of integrated circuits. It is associated with a microprocessor and an expansion bus and is bundled with the various microcomputer architectures. It plays an essential role in the computer's performance. The manufacturers were most often those who built microprocessors (AMD, Intel, etc.), but there were other precursor companies (Chips & Technologies, VIA Technologies, ZyMOS, etc.) Three generations can be identified.

The first generation, around the year 1990, was a first step in the effort to integrate everything by bringing together the glue logic and the various memory and I/O controllers using the technology of the time. It began with the 80286 and continued into the 80386 and 80486 microprocessors. It used an ISA (Industry Standard Architecture) expansion bus. The term "chipset" was fully realized with, in general, at least four VLSI (Very Large-Scale Integration) chips. The circuits were made under license from Intel.

An example was the POACH (PC-On-A-Chip) family, the ASIC (Application-Specific Integrated Circuit) using CHMOS (Complementary High-density MOS (Metal-Oxide Semiconductor)) from ZyMOS (announced at the end of 1985), with three 84-pin packages.

The 82C230 circuit (POACH1) primarily integrated the 82284 clock signal generator, the two 8259 PICs, the MC6818 (BB-)RTC ((Battery-Backed) Real-Time Clock)/ CMOS (Complementary MOS) memory[5], the 82288 bus controller and the glue logic associated with the bus.

The 82C231 circuit (POACH2) primarily integrated two 8237 DMACs (Direct Memory Access Controllers), an 8284 clock signal generator, an 8254 clock/timer, the refresh and error control logic for dynamic RAM and logic connected to I/O. These two components replaced 12 LSI components.

The 82C232 circuit (POACH-3) integrated additional discrete logic connected to buffering (*cf.* § 3.4.1 in Darche (2004)) of addresses and data, equivalent to seven LSI components. A use case for this set is the CCAT (Circuit Cellar AT) described in Ciarcia (1987a, 1987b).

5 Static RAM supplied by an autonomous energy source (battery or supercapacitor) is also called BBSRAM (Battery-Backed SRAM) or NV(S)RAM (Non-Volatile SRAM). This should not be confused with NOVORAM (NOn-VOlatile RAM, *cf.* § 4.3.2 in Darche (2012)).

Chips and Technologies (C&T) is another company that made a mark in this area with its chip sets. Among others, it produced the 82C206, a component that was part of the CS8221 CHIPSet™ called the New Enhanced AT (NEAT™). The 82C206 integrated the I/O controllers (Integrated Peripherals Controller) for DMA transfer, interrupts and timing (timer and RTC), which was missing from the previous CS8220 chipset. The final chipset was named the 82C836 Single-Chip AT (SCAT™), which integrated the preceding components.

In November 1992, Intel announced the 8240 TX/EX/ZX chipset for the 486 family under the name PCIset, with the code names of Saturn, Aries and Saturn II, in which the first intended for this expansion bus. It consisted of three components: the 82424TX Cache and DRAM Controller (CDC), the 82423TX Data Path Unit (DPU) and the 823781IB System I/O or SIO (Figure 3.16).

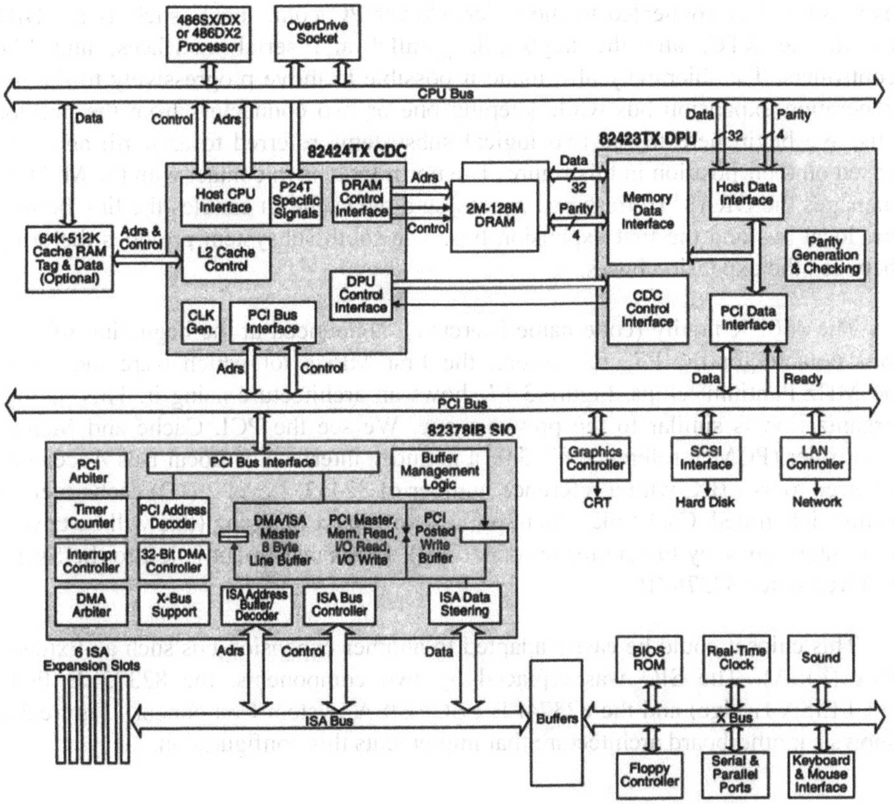

Figure 3.16. *Architecture based on the 82420 chipset for use with the 80486 (Slater 1992) © The Linley Group*

The DPU connected the host (i.e. MPU bus) and PCI buses. The SIO enabled the connection between the original ISA (Industry Standard Architecture) bus to a generic bus with a higher throughput; in other words, one that was independent of the MPU, called the PCI, for Peripheral Component Interconnect (PCI-SIG 1998, *cf.* § V2-4.2.4). Version 1.0 of the specification is dated June 1992, and version 2.0 is from April 1994. With this second generation of chipsets, we can see motherboard architecture transitioning from ISA to PCI expansion buses. This bus hierarchy (*cf.* § V2-4.1) makes it possible to adapt to the throughputs of the various subsystems.

The host bus is the fast bus because it is the bus for the MPU, as well as the cache and DRAM subsystem. The PCI expansion bus receives the fastest I/O cards, which are the video adapter, the SCSI (Small Computer Systems Interface, *cf.* § 9.3.1 in Darche (2003)) and the network interface. The X-bus connected to the ISA expansion slots connected to the older, slower PC components such as the BIOS ROM, the RTC, and the keyboard, parallel and serial interfaces, and FDD controllers. This hierarchy also made it possible to move progressively to the next generation expansion bus while keeping one or two connectors from the previous bus. We begin here to see two logical subsystems referred to as north and south based on their position in this figure. The north logic is in contact with the MPU and manages the DRAM (refresh and access) and the cache. It enables the link between the local bus and the first expansion bus. The south subsystem provides the bridge between the expansion buses.

The 430LX family (code name Mercury), announced at the beginning of 1993, was notable for the P5 architecture, the first MPUs for which were the 60 and 66 MHz Pentium[6] chips. Figure 3.17 shows an architecture using it. This family's organization is similar to the previous one. We see the PCI, Cache and Memory Controller (PCMC, reference 82434), a memory interface or Local Bus Accelerator referred to as LBX, with a reference number of 82433. Level 2 (L2) cache memory is now integrated. Cache placement policy uses direct mapping (this will be covered in a future book by the author on memories). I/O management is left to the System I/O (reference 82378ZB).

This chipset could be easily adapted to another expansion bus such as Extended ISA (EISA). The SIO was replaced by two components, the 82375EB PCEB (PCI/EISA Bridge) and the 82374EB ESC (EISA System Component). Figure 3.18 shows a motherboard architecture that implements this configuration.

6 The expected number 586 could not be used as a trademark.

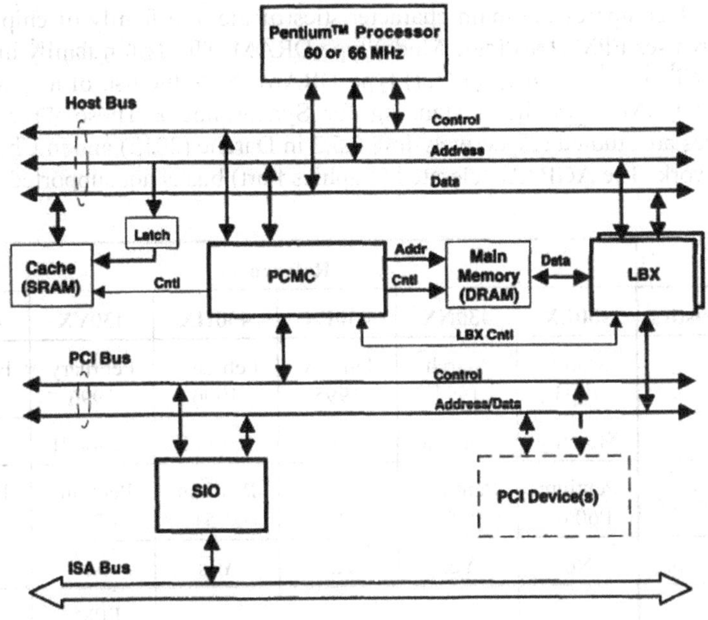

Figure 3.17. *Architecture based on the 82430 chipset for use with the Pentium (Intel 1993)*

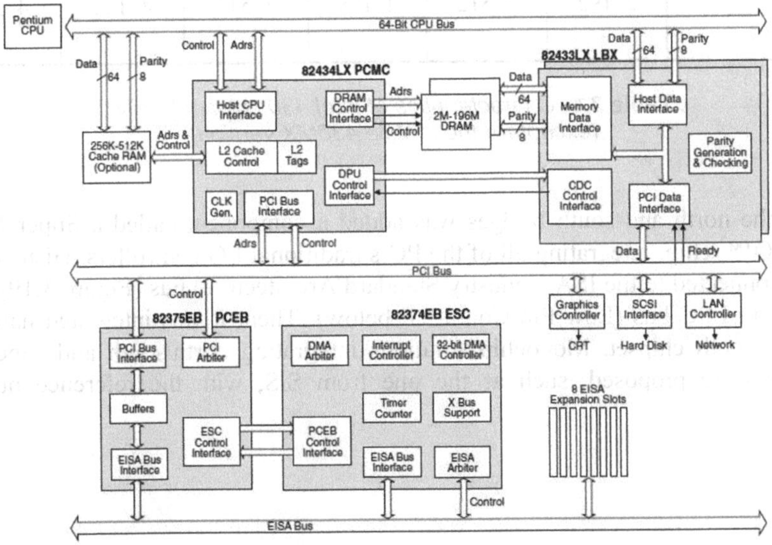

Figure 3.18. *Components in the 82430 chipset managing the EISA (Gwennap 1993) © The Linley Group*

Table 3.1 compares the main characteristics of the 430 family of chipsets. The 430LX only uses FPM (Fast Page Mode) type DRAM. The Triton family introduced the use of EDO (Extended Data Out) type DRAM. Note the use of a synchronous version (SDRAM, with the S standing for Synchronous). These three memory architectures are studied respectively in § 5.5.2 in Darche (2012) and in Chapter 6 of this same work. The AGP (Accelerated Graphics Port) bus is not supported.

	References					
Characteristics	**430LX**	**430NX**	**430FX**	**430HX**	**430VX**	**430TX**
Year	March 1993	March 1994	January 1995	February 1996	February 1996	February 1997
Code name	Mercury	Neptune	Triton	Triton II	Triton II	-
Microprocessor	Pentium P60/66	Pentium P75+	Pentium P75+	Pentium P75+	Pentium P75+	Pentium P75+
Multiprocessing	No	Yes	No	Yes	No	No
Memory type	FPM	FPM	FPM, EDO	FPM, EDO	FPM, EDO, SDRAM	FPM, EDO, SDRAM
Address capacity (MiB)	2–192	2–512	4–128	4–512	4–128	4–256

Table 3.1. *Characteristics of Intel 430 chipset family (except for the mobile 430MX version)*

To the north and south bridges was added a component called a Super I/O or Ultra I/O™[7] chip, integrating all of the PC's traditional I/O controllers. Historically, it was connected to the ISA (Industry Standard Architecture) bus (Figure 3.19), then later to an LPC bus (Low Pin Count, *cf.* below). Then, it was integrated naturally into the south chipset. Monochip solutions integrating north/south and super I/O chipsets were proposed, such as the one from SiS, with the reference number Sis 630/73x.

7 Standard MicroSystems Corporation (SMSC).

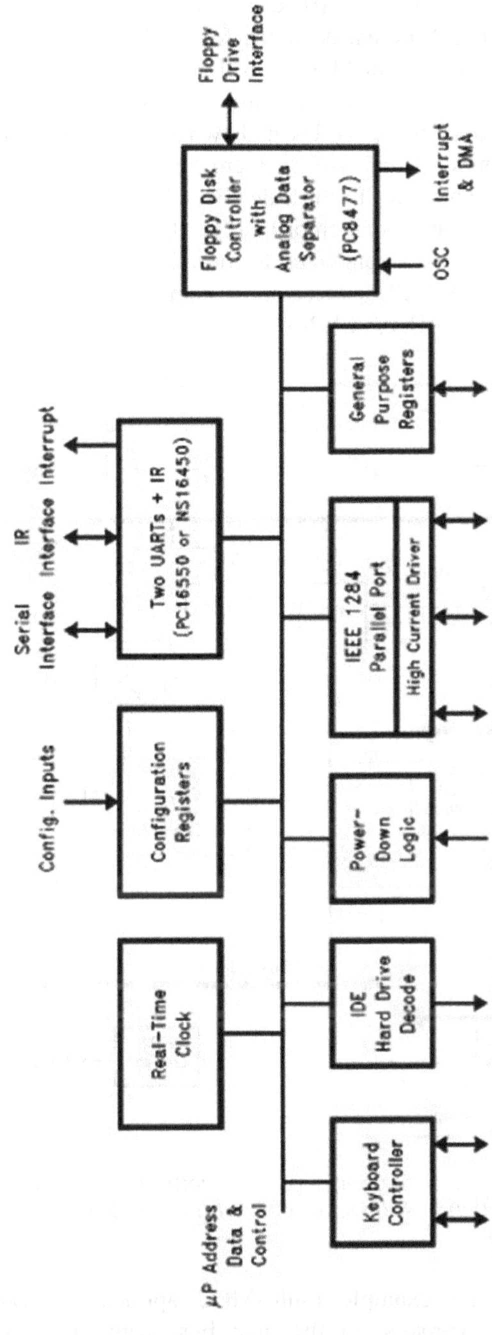

Figure 3.19. *Internal overview of a super I/O component: the PC87306 from NS*

With the Pentium Pro, II and III came the 440 and 450 families. The 440FX PCIset, code named Natoma, designed for the P6 architecture (Pentium Pro and Pentium II), was announced in May 1996. Before that, in November 1995, the 450KX/GX (respective code names Mars and Orion) for the Pentium Pro were released. As an aside, these three MPUs had two external and independent local buses, also called system buses, which were referred to respectively as the Back-Side Bus (BSB) and the Front-Side Bus (FSB). The BSB linked the microprocessor to the cache-type memory buffer (*cf.* a future sequel to Darche (2012)). The FSB or host bus is connected to the HUB or bridge, which is the real communications hub between all of these subsystems. Intel called this architecture the Dual Independent Bus (DIB, *cf.* § V2-4.2).

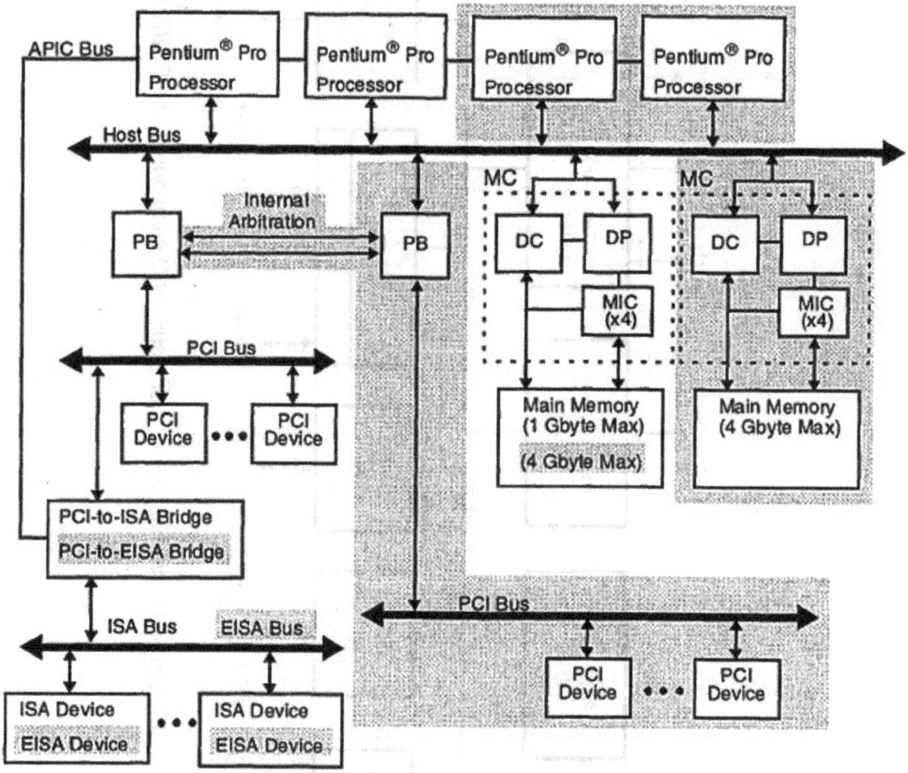

Figure 3.20. *Organization of a motherboard with an Intel 450KX/GX chipset (Intel 1996)*

Figure 3.20 shows an example multi-MPU application. The KX version supported up to two processors on the host bus, with one dynamic Memory

Controller (MC) and one host-to-PCI Bridge (PB) that connected the host bus to the PCI expansion bus. The GX version could manage up to four processors, two memory controllers and two PCI buses (grayed-out section indicates the difference between the two versions). The MC included an 82453KX DRAM Controller (DC), an 82452KX Data Path (DP) with four 82451 Memory Interface Components (MIC) per MC.

Version 2.0 of the 440LX/EX/BX added AGP support for the first time, as did the ZX a few months later in the same year. Figure 3.21 shows an example implementation. The 440 EX is a low-cost version of the 440LX designed for the Celeron family.

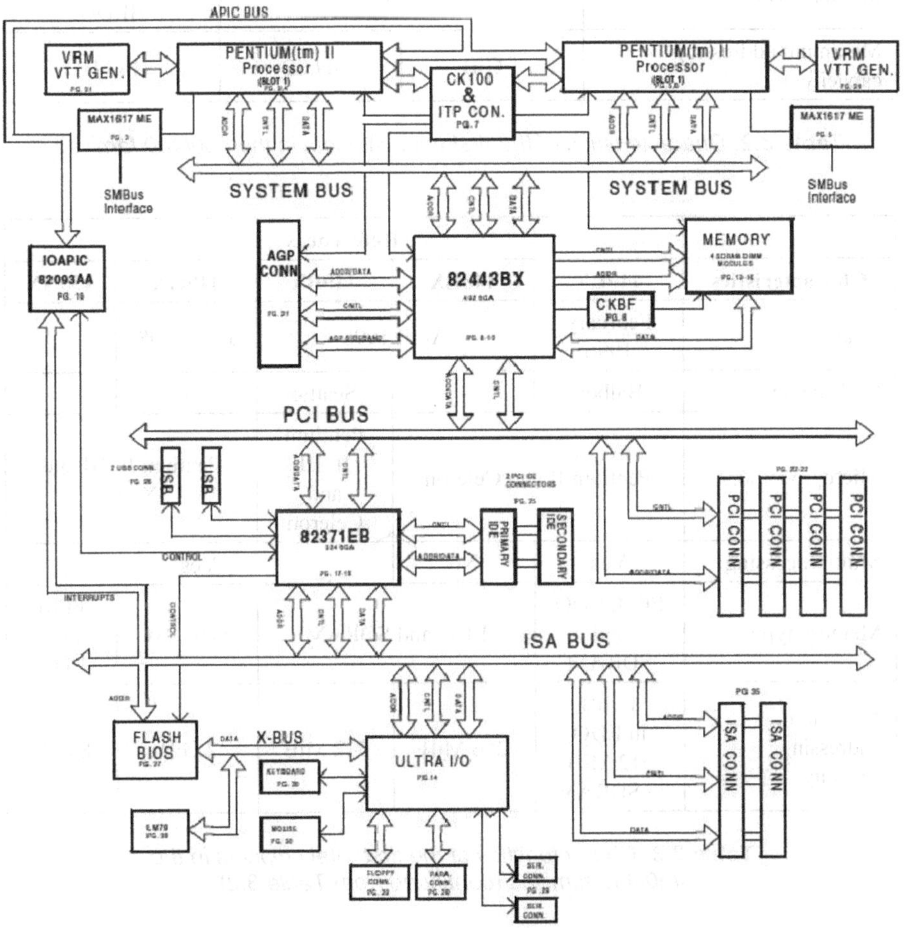

Figure 3.21. *Implementation of AGP with the 440BX chipset (source: Intel)*

Tables 3.2 and 3.3 recap the primary characteristics of the first chipsets for the 44x and 45x families.

	References		
Characteristics	**450KX**	**450GX**	**440FX**
Year	November 1995		May 1996
Code name	Mars	Orion	Natoma
Microprocessor	Pentium Pro		Pentium Pro and II
Multiprocessing	Yes		
Memory type	FPM		FPM, EDO and BEDO
Maximum addressing capacity	1 GiB	8 GiB	1 GiB

Table 3.2. *Characteristics of the first Intel chipsets in the 450/440 families*

	References				
Characteristics	**440LX**	**440EX**	**440BX**	**440GX**	**450NX**
Year	February 1996	April 1998		June 1998	
Code name	Balboa	-	Seattle	-	-
Microprocessor	Pentium II and Celeron		Pentium II, III and Celeron	Pentium II, III and Xeon	
Multiprocessing	Yes	No	Yes		
Memory type	FPM, EDO and SDRAM	EDO and SDRAM		SDRAM	FPM and EDO
Maximum addressing capacity	1 GiB in EDO 512 MiB in SDRAM	256 MiB	512 MiB	2 GiB	8 GiB

Table 3.3. *Characteristics of the first Intel chipsets in the 450/440 families (continued from Table 3.2)*

With the third generation, the meaning of the term "chip set" evolved. It now refers to the component and not to the set to which it belonged. The number of chips in the chipset was reduced to 2 + 1. We now speak about north and south chipsets. Intel introduced the Intel Hub Architecture (HUB IHA), also known as the Accelerated Hub Architecture (AHA) under the i8xx reference number, with the first being the i810 in 1999 for use with the Pentium III and Celeron chips in a socket 370 (PGA370 socket), referred to by the code name Whitney. The chipsets or bridges, the north bridge and south bridge, were respectively called the Memory Controller Hub (MCH) and the I/O Controller Hub (ICH), as shown in Figure 3.22.

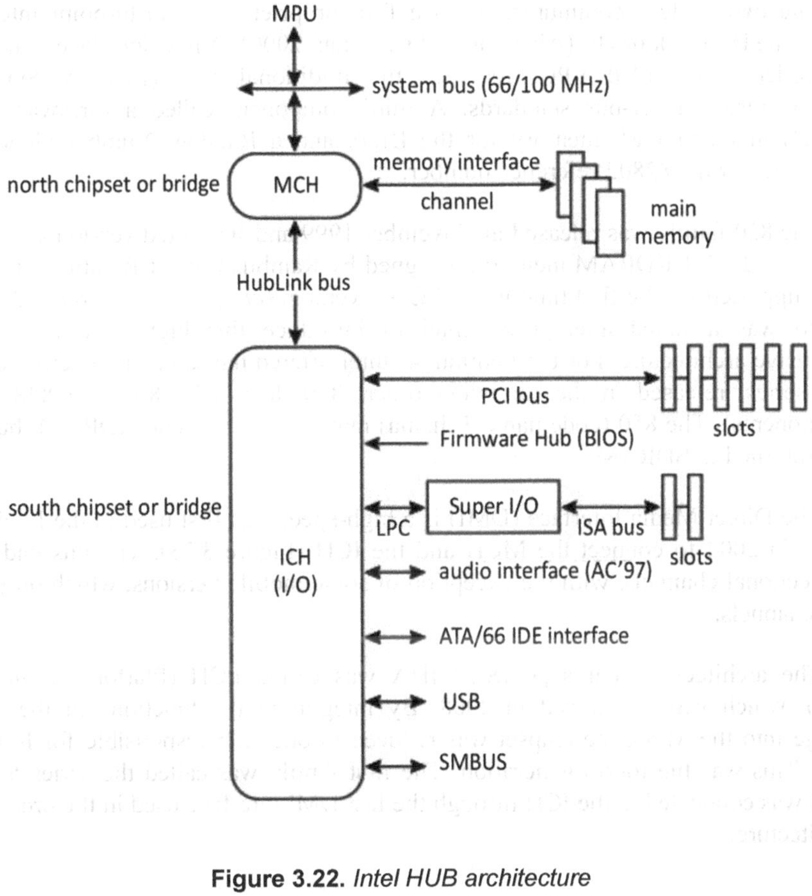

Figure 3.22. *Intel HUB architecture*

It should be noted that a graphics controller could be integrated (IGP for Integrated Graphics Processor) in the north bridge, which made it possible to manage a single system/graphics memory area called the Frame Buffer (FB). It was referred to using the acronym GMCH for Graphics and Memory Controller HUB.

Main memory was therefore partitioned, statically or dynamically, in order to share it between the microprocessor and the display controller. This is what was called the Unified Memory Architecture (UMA, will be covered in a future book by the author on memories).

The two bridges communicate via a fast, proprietary, point-to-point interface called the HubLink or HL (Ajanovic and Harriman 2000). A new interface in the I/O controller hub, called LPC, replaced the traditional ISA (Industry Standard Architecture) and X-bus standards. A third component called a Firmware Hub (FWH) included flash memory for the BIOS and a Random Number Generator (RNG) under the 82802 reference number.

The 820 family was released in November 1999 and supported version 4.0 of the AGP standard. DRDRAM memory, designed by Rambus (Direct Rambus DRAM), was supported for the first time in the PC800 version (*cf.* § 7.2.1 in Darche (2012)), which was a failed attempt by Intel to introduce this high-performance but expensive architecture. For the Pentium 4, Intel offered the same architecture in the 8xx series, released in the following order: 850, 845, 875, 865 and 848 (first components). The 850 (code name Tehama) managed PC600/800 RDRAM, but the 845 returned to SDRAM.

The Direct Media Interface (DMI) is a high-speed bus first used in the Intel 9xx series in 2004 to connect the MCH and the ICH (Figure 3.23). The bus had four bidirectional channels, with the exception of some mobile versions, which only had two channels.

The architecture that superseded IHA was called PCH (Platform Controller Hub), which was introduced in 2008. By integrating the functions of the north bridge into the MPU, the chipset was reduced to one chip responsible for I/O, the ICH. This was the fourth generation. The first family was called the series 5. The PCH was connected to the ICH through the fast DMI interface used in the preceding architecture.

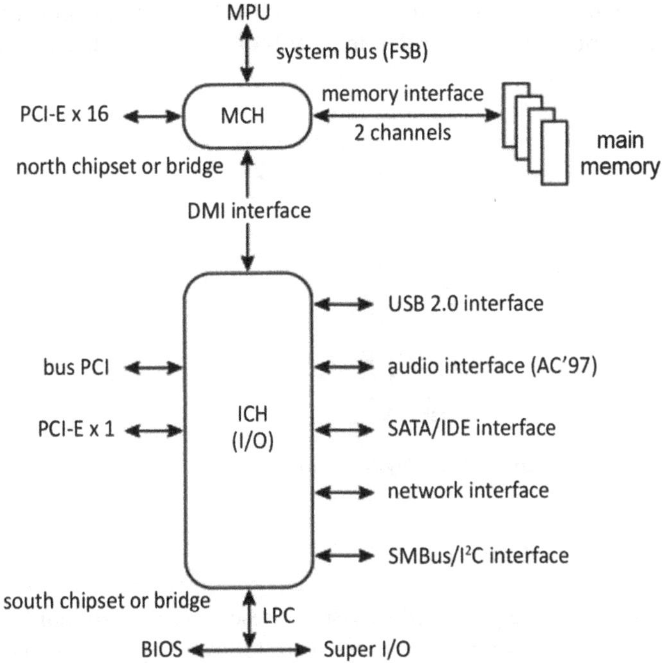

Figure 3.23. *DMI link connecting the MCH to the ICH: the 915P*

AMD used a similar technology called LDT (Lightning Data Transport), which was renamed HyperTransport (HT) in 2001 and subsequently managed by a consortium of the same name. Among others, the founding members included Apple Computer, Broadcom Corporation, NVIDIA and Sun Microsystems. LDT was a fast, bidirectional, point-to-point connection called a link between integrated circuits, which made it possible, for example, to connect an MPU to its I/O hub (i.e. the south chipset) via a component called a tunnel that contained a bridge, as shown in Figure 3.24, allowing for communication with the PCI-E bus. In addition to the bridge with tunnel, there were three other configurations – simple link, tunnel and tunnel-less bridge. The link format varied from 2 to 32 bits (growth factors of 2). The electronics used differential logic called Low-Voltage Differential Signaling (LVDS, *cf.* § 2.8 (Darche 2004) and § 3.4 (Darche 2012)), in a 1.2V version. This technology was notably used by AMD in its 64-bit MPUs (Opteron and Athlon). For more information about this bus, refer to § V2-4.4 and (Trodden and Anderson 2003; Holden *et al.* 2008).

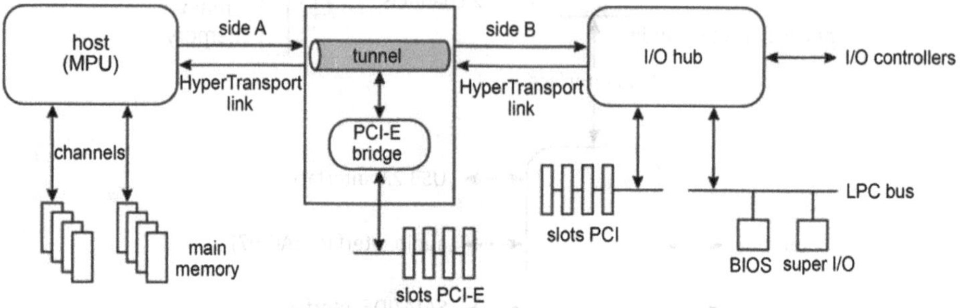

Figure 3.24. *HyperTransport Tunnel*

3.4. Motherboard architectures

The earliest models were produced by IBM. These were the original PC, the XT, and the AT, which used discrete logic. These were followed by the models described in this section. Different motherboard generations followed the chipsets (*cf.* § 3.3), which followed the MPUs, with variations depending on the Integrated Circuit (IC) package, which was housed in a support called a socket.

3.4.1. *Form factors*

In the world of PC compatibles, form factor refers to the mechanical characteristics of the motherboards in the first three IBM microcomputers (PC, XT and AT). Later, variants appeared (Figure 3.25). For example, the "full AT" became the Baby-AT, and then the ATX (Advanced Technology eXtended), which had the same form factor as the Baby-AT but was turned 90°. We should also mention the LPX (Low Profile eXtended or eXtension), NLX (New Low profile eXtended) and WTX (WorkStation eXtended) form factors. The BTX (Balanced Technology eXtended) form factor was released in September 2003. These form factors also apply to the motherboard power supply and to the outer case.

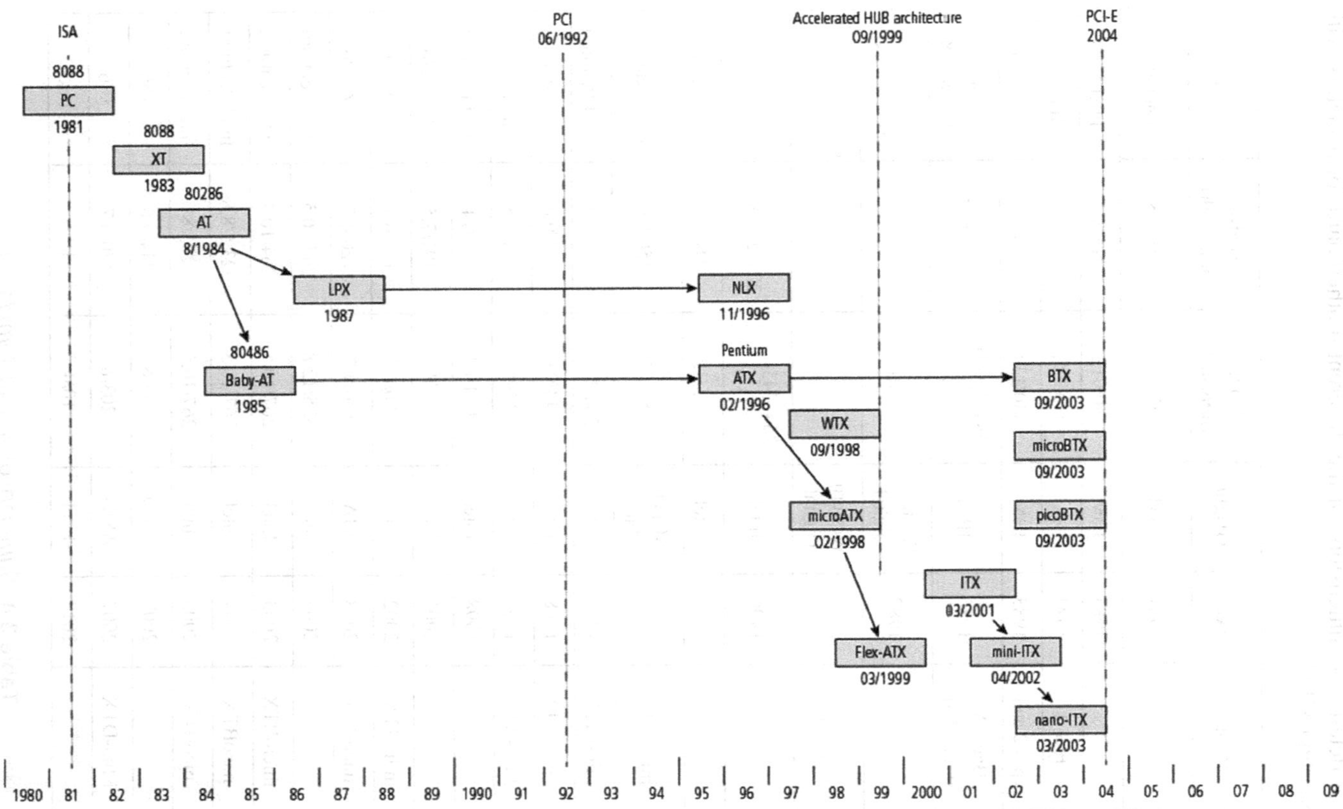

Figure 3.25. *Timeline of the main motherboard form factors*

Table 3.4 details the dimensions of the different motherboard form factors with the corresponding MPUs.

Models	Ranges	Year	Origin	Length (mm/inch)	Width (mm/inch)	MPU
Original PC (5150)		1981	IBM	279/11	216/8.5	Intel 8088
XT (5160)		1983	IBM	305/12	216/8.5	Intel 8088
AT (5170)	type 1	1984	IBM	350/13.8	305/12	Intel 286
	types 2 and 3	1985	IBM	350/13.8	236/9.3	Intel 286
	Baby-AT (XT-286)	1985	IBM	254–330/10–13	217/8.57	Intel 286/386/486
LPX		1987	Western Digital	330/13	229/9	Intel 286/386/486
	mini-LPX		Western Digital			
NLX		1996	Intel	254–345/10–13.6	203–229/8–9	
ATX		1996	Intel	305/12	244/9.6	Intel Pentium
	mini-ATX		AOpen Inc.	150/5.9	150/5.9	Intel Pentium
	mini-ATX	1997	Intel	284/11.2	208/8.2	Intel Pentium
	microATX	1998	Intel	244/9.6	244/9.6	Intel Pentium
	Flex-ATX	1999	Intel	229/9	191/7.5	Intel Pentium
WTX		1998	Intel	425/16.75	356/14	Intel Pentium
ITX		2001	VIA	215/8.5	191/7.5	VIA/x86
	mini-ITX	2002	VIA	170/6.7	170/6.7	VIA/x86
	nano-ITX	2003	VIA	120/4.7	120/4.7	VIA/x86
BTX		2003	Intel	325/12.8	267/10.5	Intel Pentium
	microBTX	2003	Intel	267/10.5	264/10.4	Intel Pentium
	nanoBTX	2004	Intel	267/10.5	223.5/8.8	Intel Pentium
	picoBTX	2003	Intel	267/10.5	203/8	Intel Pentium
DTX		2007	AMD	203/8	244/9.6	x86
	Mini-DTX	2007	AMD	203/8	170/6.7	x86
NUC		2012	Intel	10/4	10/4	Intel

Table 3.4. *Summary of various form factors*

3.4.2. *Current motherboard architecture*

Figure 3.26 shows the motherboard for a contemporary microcomputer with the microATX form factor. With a few exceptions such as the power supply connectors, the components are all surface mounted (SMD for Surface-Mounted Device), which is associated with SMT (Surface Mount Technology). The main component is the microprocessor, here a Core™. Only its Zero Insertion Force (ZIF) socket is shown. The second subsystem is the RAM, made up of four 2400 MHz DDR4 (*Double Data Rate*, *cf.* § 6.5 (Darche 2012)) DIMMs (Dual-In-line Memory Modules), inserted into memory slots. In comparison to the first PC, what is remarkable is the disappearance of the glue logic that enabled management of and communication with memory; the I/O exchange units have also disappeared in favor of a single B250 south chipset, with the north chipset being integrated into the MPU. However, it is surrounded by numerous components that supply the MPU with power (radial capacitors and coils). The GPU (Graphics Processing Unit) is integrated into the MPU.

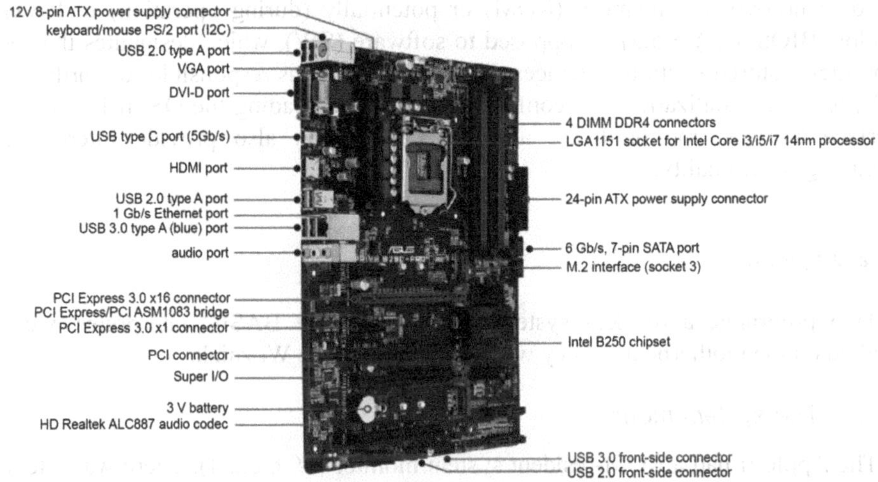

Figure 3.26. *Current PRIME B250-PRO motherboard (source: Asus – 2018, completed). For a color version of this figure, see www.iste.co.uk/darche/microprocessor5.zip*

The I/O connectors are found on the side of the motherboard that touches the back of the computer case. It connects to external peripherals. Other connectors distributed on the other sides make it possible to either connect a mass storage unit or complement the rear connectors by allowing them to be moved to the front or rear side of the case.

3.5. Evolution of microcomputer firmware

At startup, all computers require a program stored in ROM called firmware. Three companies have made their mark with their firmware: Apple, IBM and Sun. They are described after a precise definition of the term.

3.5.1. *Definition*

Historically, the term is attributed to Acher Opler from an article in *Datamation* magazine in January 1967 (Opler 1967). Firmware (FW) or micrologic is the low-level logic that controls a microprocessor-based system, at least on startup. It is stored in non-volatile memory (ROM) or potentially (during operation) in RAM (shadow BIOS, *cf.* § 3.5.3), as opposed to software (SW), which designates the set of programs stored in storage devices. For a computer, it is responsible for hardware (HW) testing, initialization and configuration, and for loading the OS. It is also an interface between the hardware and the OS. It can also provide interactive debugging functionality.

3.5.2. *Apple II*

Two programs, a resident system monitor and the BASIC interpreter, were installed on the motherboard. They were written by Steve Wozniak.

3.5.2.1. *The system monitor*

The Apple II had a ROM-resident system monitor (*cf.* § 2.2.4). There was a text editor and a so-called self-prompting command interpreter; in other words, it waited for input of one of 22 provided commands in the form of a letter. This firmware made it possible to read and disassemble memory, to assemble (mini-assembler) a program, to write, move or compare a block, and to load and launch execution of a program at a given address. It also enabled reading and writing from a cassette tape. Debugging commands made it possible to examine and modify the MPU 6502's registers, among other actions.

3.5.2.2. *BASIC ROM*

The firmware also included a BASIC language interpreter that was launched either manually or automatically depending on the version (Autostart ROM).

3.5.3. *PC BIOS*

In the early days of computing, applications directly managed input–output and used specialized libraries. The invention of operating systems allowed them to offload this management. The advent of the microcomputer with the CP/M operating system (Control Program for Microcomputers from Digital Research) introduced the notion of a BIOS, which allowed the OS, which was called BDOS (Basic Disk Operating System), to rely on a software layer for input/output management (Figure 3.27a).

The BIOS was primarily designed for disk management. Services were launched by calls from subprograms. The PC with DOS (Disk Operating System) popularized this principle. However, calling these routines was done by software or hardware interrupt (*cf.* § V4-5.1), and the processor's registers were used to pass incoming and outgoing parameters. Originally kept in secondary memory for CP/M, the BIOS for the PC was moved to ROM on the motherboard. The BIOS (Basic Input/Output System) is therefore software stored in ROM (firmware). It dates back to the first PC (1981). It was difficult to build for because it had never been officially specified; the only reference document was IBM's hardware reference manual (IBM 1981, 1984). It was specifically designed for the x86 architecture (processors and related circuits – I/O controllers). At initialization, the MPU connected to the reset vector (= $FFF0_{16}$) address. An unconditional jump instruction labeled RESET was stored there, at the beginning of the BIOS with the POST program.

This is why BIOS is found at the top of the address space (Figure 3.27b). It executes in real-time, that is, there is no management of virtual memory, and the memory area, which is limited to 640 KiB (!), is not protected. As a result of the limitations of this approach, an application can directly use a BIOS routine by bypassing the OS, and can even directly manage hardware (in the case of game software). Indeed, the two previously mentioned OSes were single task/single user, and an application could indefinitely lock up the computer, for example by hiding interrupts.

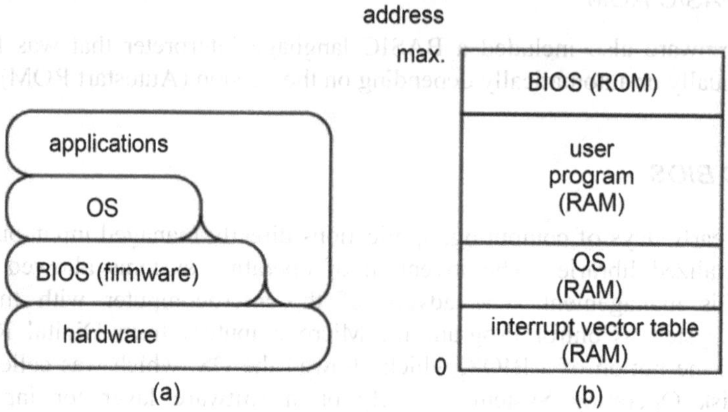

Figure 3.27. *The different functional layers of an old-generation PC microcomputer and a diagram of main memory for the original PC (1981)*

The BIOS provides three essential functions: Power-On Self-Test (POST), hardware initialization and first-level loading. It also provides an interface between the OS and the hardware in the form of Interrupt Service Routines (ISR). The final program, for configuration (setup), is not present on every machine.

The POST for the original PC ran a hardware diagnostic to detect malfunctions. At startup, it tested the MPU, the main memory, the BIOS and the DRAM. The MPU test examined flags and associated conditional jumps, as well as the other registers. The test for the ROM that contained the BIOS was based on a checksum (*cf.* III.6.4 in Darche (2000)). The DMA controller was then tested and initialized in order to refresh the DRAM. The first 32 KiB of DRAM was then tested by reading and writing a 5-byte sequence (base 16) that consisted of AA, 55, FF, 01 and 00. The 8259 interrupt controller was then initialized, as was the interrupt vector table (*cf.* § V4-5.7). The video controller was initialized. The IT controller was then tested. The keyboard was tested. Testing the RAM above 32 KiB then continued by calculating its total size. There could also be additional BIOS at the interface-board level (range of $C8000{:}F4000_{16}$ by 2 KiB block), whose presence was indicated by signing the first two bytes $55AA_{16}$. It was detected and then tested. The program determined whether a floppy disk drive was attached and, if so, loaded the primary bootstrap loader found on the disk. Note that in modern computers, the primary boot loader is found in general in the first sector (no. 1) of the first track (no. 0), called the MBR (Master Boot Record) on the primary hard drive's active partition. The boot sequence consists of finding the OS, loading it and transferring control to it. The function of the primary boot loader is to find the secondary boot(strap) loader, which will then finally load the operating system.

The setup program specifically allowed the date and time to be adjusted, as well as the boot priority order for the mass storage devices. For more details on the BIOS, see, for example, Croucher (2004).

Since ROM access time[8] is slower than RAM access time, the BIOS can be transferred when the machine is initialized to improve execution time for internal routines. This technique is called shadow BIOS. A microcomputer can have a dual BIOS, referred to as the primary and secondary BIOS. During the initialization sequence, the primary BIOS is tested, for example using a checksum or a Cyclic Redundancy Check (CRC), described respectively in § III.6.4 and III.6.7 in Darche (2000); also *cf.* § 2.6.4 in Darche (2012). If the test is negative, the secondary BIOS is selected and tested. If this test is negative, the machine is halted. This technique makes it possible, among other things, to recover the machine in the case of a failed firmware update. The flow chart is presented in Noll (1998).

The BIOS is fixed and proprietary. The binary image, once generated, cannot be changed. It is executed in real-time mode with an address limitation of 1 MiB. To resolve this situation, Intel proposed a new version called EFI (Extensible Firmware Interface), which is described in Intel (2000), initially called the Intel Boot Initiative or IBI (1997) and designed for the Itanium architecture. This was a standardized and independent interface between the motherboard's firmware and the OS (Figure 3.28) that also provided an Application Programming Interface (API). Thus, the OS was allowed to launch and take charge of the underlying hardware such as mass storage, network or display interfaces. It was written in a modular fashion in a High-Level programming Language (HLL), the C language being most suitable, and not in assembly language, as is the BIOS, to facilitate development, maintenance, updating and portability.

This architecture provides a shell for launching setup and diagnostic utilities. The initialization phase for the BIOS remained unchanged. It managed all of the mass storage devices as well as the network interfaces. It also had an expansion mechanism that allowed it to adapt to new peripherals. It also managed drivers.

---

8 Access time for RAM or ROM is defined as the time elapsed between presentation of the address and data access, with the operation, in this case a read, being subsequently carried out. For more details, *cf.* § 1.1 and 3.3 in Darche (2012).

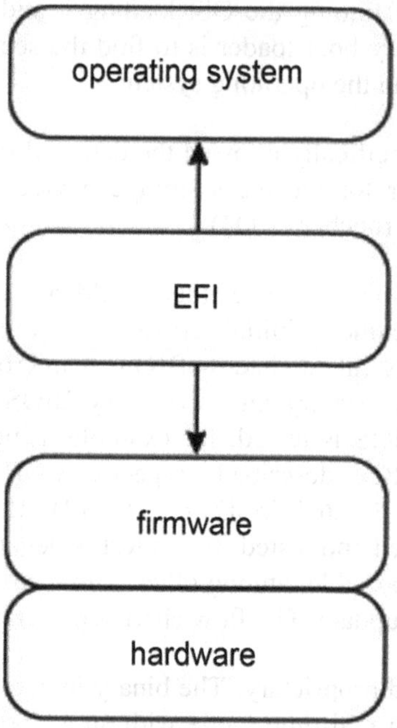

Figure 3.28. *EFI software location*

Unified EFI (UEFI) succeeded EFI and is supported by an association founded in 2005, whose members include several hundred companies, including Intel and Microsoft. It supports non-Intel architectures such as Arm®. Its specification is almost 3,000 pages long (UEFI 2017).

Originally, the BIOS could not be updated because it was stored in ROM. EPROM (Erasable PROM) technology made it possible to program the ROM electrically using an external programmer, with erasing being done using ultraviolet light (for more on this subject, *cf.* § 2.1.2). Today, non-volatile memory (NVM) is programmable and erasable electrically (Flash EEPROM/E^{2}PROM for Electrically EPROM or FEEPROM/FE^{2}PROM). Programming is carried out using a memory dump in the form of a file stored locally, originally on a disk and later on a local or even a remote storage device. Firmware for the most well-known PCs came from AMI (American Megatrends, Inc.) and Phoenix Technologies.

Additional information is found in Dice (2018).

ADDITIONAL INFORMATION.– BSP (Board Support Package) is low-level firmware designed for an embedded system (the original source of the term). It sits between the hardware and the OS. It provides hardware abstraction for the OS by converting its generic function calls into I/O routines. Its functions, in order of definition, are initialization, boot, hardware management (drivers) and interrupt management. It may contain information files concerning the board hardware. The IBM PC BIOS can be seen as its forerunner.

3.5.4. *Open firmware*

Sun workstations had a monitor residing in PROM (Programmable Read-Only Memory) called open firmware, also called OpenBoot, originally developed by Sun. This firmware is hardware independent. It was standardized by the IEEE under reference number IEEE Std 1276-1994 (IEEE 1994a, 1994b, 1994c, 1995). It enabled subsystem diagnostics (POST) on startup, including the keyboard, main memory, the mass storage devices, the I/O controllers, etc. It could also be launched during startup of the computer to allow for management commands to be entered from an interpreter. It was initially used in Sun workstations designed around an MPU from the SPARC (Scalable Processor ARChitecture) family and in PowerPC-based Macintosh computers.

3.6. Conclusion

We have described the architecture of the first emblematic microcomputers, the Apple II and the various generations of the PC. The latter were architected around a chipset. These chipsets were then presented. What this chapter has shown is the progressive integration of discrete logic components, and then subsystems, to arrive at a single component that is peripheral to the MPU and main memory. Then, to conclude the chapter, the firmware aspect was addressed by presenting the PC's BIOS and (U)EFI.

Conclusion of Volume 5

The microprocessor lies at the heart of current digital systems. This programmable logic component sequentially executes instructions from a program stored in main memory. The preceding volume addressed the software aspects of this component.

In its opening chapter, this final volume presented the software development chain for a microprocessor system application. The various steps and related tools were examined.

The software and hardware tools for debugging and testing were the subject of the second chapter. Early debugging tools and practices such as *postmortem* analysis and the iterative burn and learn approach are no longer suitable today. Increasingly complex and fast MPUs require integration of testing and debugging logic as close to operations as possible. Internal debugging (OCD for On-Circuit Debugger/Debugging) and the related communication interfaces such as JTAG (Joint Test Action Group) were then examined. Assembly language, which enables the symbolic manipulation of instructions and variables, was then described in detail. Even though today, High-Level (programming) Languages (HLL) have taken its place in embedded systems thanks to progress in storage and compilers, it is still used in specific cases such as memory and runtime optimization, to interface with the system or hardware, and in teaching computer architecture.

To conclude this final volume, the architectures of the early emblematic microcomputers, that is, the Apple II and the various generations of the PC, were presented. The integration of discrete logic components and then the microcomputer's subsystems into a chipset occurred progressively, arriving today at a single external component that, in combination with the MPU and main memory, forms an almost complete processing unit. Finally, the firmware aspect was addressed by presenting the BIOS (Basic Input/Output System) and the (U)EFI ((Unified) Extensible Firmware Interface) for the PC (Personal Computer).

A subsequent work will address the next generation, which corresponds approximately to the 1980s and 1990s. This will include 16/32-bit and RISC-type (Reduced Instruction Set Computer) microprocessors. Two important concepts relating to memory were introduced in these systems, namely virtual and cache memory. Another major aspect introduced in these processors was Instruction-Level Parallelism (ILP) with pipelined and superscalar architectures. The final book will discuss multicore technology.

NOTE.– The concepts introduced in this work will be completed in relation to newer releases in future works by the author.

Exercises

This exercise supplements the ideas presented in this work. Their numbering refers to the chapter to which they are related.

Chapter 1. Exercises

E1.1. What is the difference between assembly language and the assembler?

Answer. Assembly language is the fundamental symbolic language for the processor, and the assembler is the program that performs, among other tasks, the translation of symbols into machine code.

Acronyms

This chapter includes all of the acronyms used in this volume. They are used after the first reference in each chapter.

General

A

A	Address
ABI	Application Binary Interface
A/D	Analog/Digital
ADC	Analog-to-Digital Converter
AGP	Accelerated Graphics Port
AHA	Accelerated Hub Architecture (Intel), *cf.* IHA
AL	Assembly Language
ALE	Address Latch Enable
AMI	Abstract Machine Interface
API	Application Programming Interface
ASIC	Application-Specific Integrated Circuit

AT	Advanced Technology
ATA	AT Attachment
ATX	Advanced Technology eXtended

B

b	bit (*cf.* BIT)
B	Binary
B	Byte
BASIC	Beginner's All-purpose Symbolic Instruction Code
BB-RTC	Battery-Backed RTC
BBSRAM	Battery-Backed SRAM (*cf.* NV(S)RAM)
BCS	Binary Compatibility Standard
BDC	Background Debug Controller (Motorola)
BDM	Background Debug Mode
BDM	Background Debug Module (Motorola)
BDOS	Basic Disk Operating System
BEDO DRAM	Burst EDO DRAM
BGA	Ball Grid Array
B(G)DM	Back(Ground) Debug Mode
BIOS	Basic Input/Output System
BIST	Built-In Self-Test
BIT	BInary digiT or Binary digIT
BM	Background Mode

BNF	Backus and Naur Form
BR	Bypass Register
BSB	Back-Side Bus
BSC	Boundary-Scan Cell
BSD	Berkeley Software Distribution
BSDL	Boundary-Scan Description Language
BSP	Board Support Package
BSR	Boundary-Scan Register
BTX	Balanced Technology eXtended
BUFFALO	Bit User Fast Friendly Aid to Logical Operation

C

CA	Conditional Assembly
CCAT	Circuit Cellar AT
CDC	Cache and DRAM Controller
CGA	Color/Graphics Adapter
CHMOS	Complementary HMOS (Intel)
CLK	CLocK
CMOS	Complementary MOS
COBOL	COmmon Business Oriented Language
COFF	Common Object File Format
CP/M	Control Program for Microcomputers (Digital Research)

CPU	Central Processing Unit
CRC	Cyclic Redundancy Check
CRT	Cathode Ray Tube

D

D	Data
D	Decimal
DAP	Debug Access Port (ARM)
DC	DRAM Controller
DCC	Debug Communication Channel (ARM)
DDR	Double Data Rate
DfT	Design for Testability
DIB	Dual Independent Bus
DIL	Dual-In-Line
DIMM	Dual-In-line Memory Module
DIP	DIL Package
DLL	Dynamic Link Library
DLX	DeLuXe
DMA	Direct Memory Access
DMAC	DMA Controller
DMI	Direct Media Interface (Intel)
DOS	Disk Operating System
DP	Data Path

DPU	Data Path Unit
DR	Debug Register
DRAM	Dynamic RAM
DRDRAM	Direct RDRAM
DSCLK	Data Serial CLocK
DSI	Data Serial Input
DSO	Data Serial Output
DSP	Digital Signal Processor
DUT	Device Under Test
DVI	Digital Visual Interface
DVI	Digital Video Interactive (obsolete)
DVI-D	DVI-Digital

E

EABI	Embedded ABI
EDC	Error-Detecting Circuit/Code
EDO	Extended Data Out
EDSAC	Electronic Delay Storage Automatic Calculator
EEMS	Enhanced Expanded Memory Specification
EEPROM	Electrically EPROM
E^2PROM	Electrically EPROM (less common)
EFI	Extensible Firmware Interface
EGA	Enhanced Graphics Adapter

EICE	Embedded ICE
EISA	Extended ISA
ELF	Executable and Linkable Format
EMM	Expanded Memory Manager
EMS	Expanded Memory Specification (LIM EMS)
EPROM	Erasable PROM
ESC	EISA System Component 82374
ETM	Embedded Trace Macrocell (ARM)

F

FD	Floppy Disk
FDD	Floppy Disk Drive
FDM	Foreground Debug Mode
FEEPROM	Flash EEPROM (JEDEC – JESD88C definition)
FE^2PROM	Flash EEPROM (less common)
F(G)DM	Fore(Ground) Debug Mode
FIFO	First In, First Out
FL	Long Floating Point (IEEE Std 694-1985)
FP	Packed decimal Floating Point (IEEE Std 694-1985)
FPGA	Field-Programmable Gate Array
FPM	Fast Page Mode
FR4	Flame Retardant 4
FS	Short Floating Point (IEEE Std 694-1985)

FSB	Front-Side Bus
FW	FirmWare
FWH	Firmware Hub (Intel)
FX	Extended Floating Point (IEEE Std 694-1985)

G

GMCH	Graphics and Memory Controller HUB
GNU	GNU's Not UNIX
GPU	Graphics Processing Unit/Graphics Processor Unit

H

H	Hexadecimal
HAL	Hardware Abstraction Layer
HD	Hard Disk
HDD	HD Drive
HDL	Hardware Description Language
HL	HubLink (Intel)
HLA	High-Level Assembly
HLL	High-Level (programming) Language
HMA	High Memory Area
HMOS	High-density MOS (Depletion-load NMOS)
HT	HyperTransport
HW	HardWare

I

iBCS	Intel BCS
IBI	Intel Boot Initiative (later EFI)
IC	Integrated Circuit
ICD	In-Circuit Debug
ICE	In-Circuit Emulator
ICH	I/O Controller HUB
ICS	In-Circuit Simulator
IDE	Integrated Drive Electronics
IDR	IDentifier Register
IGP	Integrated Graphics Processor
IHA	Intel Hub Architecture (synonym of AHA)
I^2C™	Inter-IC
ILP	Instruction-Level Parallelism
INT	Interrupt (*cf.* IT)
I/O	Input/Output
IO	Input/Output (rarely used)
IoT	Internet of Things
IR	InfraRed
IRQ	Interrupt Request
ISA	Instruction Set Architecture
ISA	Industry Standard Architecture
ISBN	International Standard Book Number

ISC In-System Configurable

ISP *In-Situ* Programming

ISR Interrupt Service Routine

IT InTerruption (*cf.* INT)

ITP In-Target Probe

J

JTAG Joint Test Action Group

L

LAN Local Area Network

LBX Local BUS Accelerator

LCD Liquid Crystal Display

LCF Linker Control File

LDT Lightning Data Transport (AMD) renamed HyperTransport

LED Light-Emitting Diode

LGA Land Grid Array

LIM Lotus Software, Intel, Microsoft

LIW Long Instruction Word

LLL Low-Level (programming) Language

LPC Low Pin Count

LPX Low Profile eXtended or eXtension (Western Digital)

LSI Large-Scale Integration

LVD	Low-Voltage Differential
LVDS	LVD Signaling

M

MASM	Microsoft Macro Assembler
MBR	Master Boot Record
MC	Memory Controller
MCH	Memory Controller HUB
MCU	MicroController Unit (preferred), *cf.* μP
MCU	MicroComputer Unit (*cf.* μC)
MDA	Monochrome Display Adapter
MDS	Microcomputer Development System (Intel MDS 80)
MEMR	MEMory Read
MEMW	MEMory Write
MIC	Memory Interface Components
MIPI	Mobile Industry Processor Interface
MOS	Metal-Oxide Semiconductor
MPSD	Modular Port Scan Device (TI)
MPU	MicroProcessor Unit (*cf.* μP)
MS-DOS	MicroSoft DOS

N

NB	Natural Binary
NBC	Natural Binary Code (*cf.* NB)

NEAT	New Enhanced AT
NLX	New Low profile eXtended
NMI	Non-Maskable Interrupt
NMOS	Negative (channel) MOS
NOP	No Operation
NOVORAM	NOn-VOlatile RAM (*cf.* NOVRAM)
NOVRAM	NOn-Volatile RAM (*cf.* NOVORAM)
NRZ	Non-Return to Zero
NTSC	National Television Standards Committee
NUC	Next Unit of Computing (Intel)
NVM	Non-Volatile Memory
NVRAM	Non-Volatile RAM (*cf.* BBSRAM)
NVSRAM	Non-Volatile SRAM (*cf.* BBSRAM)

O

OC	Output Capture
OCD	On-Circuit Debugger/Debugging
OCDS	On-Chip Debug System
OLE	Object Linking and Embedding (Microsoft)
OMF	Object Module Format (Intel)
OnCE	On-Chip Emulation (Motorola)

OS	Operating System
OTP	One-Time Programmable
OTPROM	One-Time EPROM

P

PA	Physical Address
PB	Host-to-PCI Bridge
PC	Personal Computer
PC/AT	Personal Computer Advanced Technology
PCB	Printed Circuit Board
PCEB	PCI/EISA Bridge
PCH	Platform Controller Hub
PCI	Peripheral Component Interconnect (standard)
PCI-E or PCIe	PCI Express
PCMC	PCI, Cache and Memory Controller (82434)
PDI	Program and Debug Interface
PE	Portable Executable
PIA	Peripheral Interface Adapter (Motorola)
PIC	Programmable Interrupt Controller
POACH	PC-On-A-Chip (Zymos)
POSIX	Portable Operating System Interface eXchange
POST	Power-On Self-Test
PPI	Programmable Peripheral Interface

PROM	Programmable ROM
PS/2	Personal System/2

Q

Q	Octal

R

RAM	Random Access Memory
RDRAM	Rambus DRAM
RDY	Ready
RF	Resume Flag
RISC	Reduced Instruction Set Computer
RMB	Reserve Memory Byte
RNG	Random Number Generator
ROM	Read-Only Memory
RS	Recommended Standard
RTC	Real-Time Clock
RTCK	Return TCK
RTOS	Real-Time OS

S

SAP	SHARE Assembly Program (MIT)
SAP	Symbolic Assembly Program (IBM 704) - original definition
SATA	Serial ATA

SCAT	Single-Chip AT
SCSI	Small Computer Systems Interface
SDI	Serial Debug Interface
SDRAM	Synchronous DRAM
SIO	System I/O
SMB/SMBus	System Management Bus
SMC	Surface Mounted Component
SMD	Surface Mounted Device
SMT	Surface Mount Technology
SOAP	Symbolic Optimizer and Assembly Program (IBM 650)
SoC	System on (a) Chip, System-on-Chip
SoM	System on Module
SPARC	Scalable Processor ARChitecture
SPI™	Serial Peripheral Interface
SR	Shift Register
SVID	System V Interface Definition
SW	SoftWare
SWD	Serial Wire Debug (ARM)
SYM-1	SIM Model 1 (Synertek Systems Corp.)

T

TAP	Test Access Port
tasm	Turbo-Assembler (Borland)

TCK	Test ClocK
TDI	Test Data Input
TDO	Test Data Output
TF	Trap Flag
THT	Through-Hole Technology
TMS	Test Mode Select
TOS	Top Of Stack
TRST	Test ReSeT
TTL	Transistor–Transistor Logic

U

UART	Universal Asynchronous Receiver Transmitter (Texas Instruments)
UEFI	Unified EFI
UMA	Unified Memory Architecture
UMA	Upper Memory Area
UMB	Upper Memory Block
USB	Universal Serial Bus
UV	UltraViolet
UV-EPROM	UltraViolet EPROM

V

VGA	Video Graphics Array
VHDL	VHSIC Hardware Description Language

VHSIC	Very High Speed Integrated Circuit
VIM-1	Versatile Input Monitor 1
VLIW	Very LIW
VLSI	Very LSI
VM	Virtual Memory

W

WBR	Wrapper Boundary Register
WBY	Wrapper BYpass register
WIR	Wrapper Instruction Register
WSC	Wrapper Serial Control
WSI	Wrapper Serial Input
WSO	Wrapper Serial Output
WTX	WorkStation eXtended

X

XMA	Expanded Memory Adapter (IBM)
XMS	eXtended Memory Specification
XT	eXtended Technology

Z

ZIF	Zero Insertion Force

Miscellaneous

μC	Microcontroller (*cf.* MCU)
μC	Microcomputer
μP	Microprocessor (*cf.* MPU)

Units of measurement and unit prefixes

G	giga (= 10^9)
gibi	gigabinary (prefix Gi)
GiB	gibibyte
K	kilo binary (= 1024) – a deprecated prefix; the capital letter indicates the value of the prefix (which we have selected for this work)
Kb	kilobit (= 1024 b) – former multiple, to be avoided
KB	kilobyte (= 1024 bytes)
Kib	kibibit
Kibi	kilobinary (prefix Ki)
KiB	kibibyte
M	mega (= 10^6)
MB	megabyte
Mebi	megabinary (prefix Mi)
MiB	mebibyte
MIPS	Million Instructions Per second

Electrical characteristics

Gnd	Ground
V_{CC}	Collector DC supply voltage
V_{DC}	DC voltage
V_{DD}	Drain DC supply voltage
V_{TT}	TerminaTion rail voltage

Company or organization

ACM	Association for Computing Machinery
AFISI	*Association Française d'Ingénierie des Systèmes d'Information*
AIEE	American Institute of Electrical Engineers
AMD	Advanced Micro Devices, Inc.
AMI	American Megatrends, Inc.
ARM	Acorn RISC Machine, later Advanced RISC Machines
AST, Inc.	Albert Wong, Safi Qureshey and Thomas Yuen
CEI	*Commission Electrotechnique Internationale* (*cf.* IEC)
C&T	Chips and Technologies
DIN	*Deutsches Institut für Normung*
EMS	Enhanced Memory Systems, Inc.
HP	Hewlett-Packard
IBM	International Business Machines Corporation
IEC	International Electrotechnical Commission (*cf.* CEI)
IEEE	Institute of Electrical and Electronics Engineers

IRE	Institute of Radio Engineers
ISO	the International Organization for Standardization
ISTO	IEEE Industry Standards and Technology Organization
JEDEC	Joint Electron Device Engineering Council (Solid-State Technology Association)
MPR	Microprocessor Report
OCP-IP	Open Core Protocol International Partnership
PCI-SIG	PCI Special Interest Group
SMSC	Standard MicroSystems Corporation
ST	Seagate Technology
TI	Texas Instruments
UCR	University of California Riverside

Trademarks (™)

I^2C	Philips
MCA	IBM
NEAT	Chips and Technologies (C&T)
OnCE	Motorola
Pentium	Intel Corporation
SCAT	Chips and Technologies (C&T)
SPI	Motorola

Registered trademarks (®)

AMD	AMD
AT&T	AT&T
Intel	Intel
Micro Channel	IBM Corporation
MIPI	Mobile Industry Processor Interface
Pentium	Intel
Rambus	Rambus, Inc.
tasm	Borland
UNIX	AT&T
Windows	Microsoft Corporation

References

For reasons of consistency in this domain, the references have been organized by chapter.

Preface

Darche, P. (2000). *Architecture des ordinateurs – Représentation des nombres et codes – Cours avec exercices corrigés*. Éditions Gaëtan Morin.

Darche, P. (2002). *Architecture des ordinateurs – Fonctions booléennes, logiques combinatoire et séquentielle – Cours avec exercices et exemples en VHDL*. Éditions Vuibert.

Darche, P. (2003). *Architecture des ordinateurs – Interfaces et périphériques – Cours avec exercices corrigés*. Éditions Vuibert.

Darche, P. (2004). *Architecture des ordinateurs – Logique booléenne: implémentations et technologies*. Éditions Vuibert.

Darche, P. (2012). *Mémoires à semi-conducteurs: principe de fonctionnement et organisation interne des mémoires vives – Volume 1*. Éditions Vuibert. One of four books selected for the AFISI (Association Française d'Ingénierie des Systèmes d'Information) prize for the best computing book 2012.

Part 1 (Chapters 1 and 2)

Agrawal, V.D., Kime, C.R., and Saluja, K.K. (1993a). A tutorial on built-in self-test. Part 1: Principles. *IEEE Design & Test of Computers*, 10(1), 73–82.

Agrawal, V.D., Kime, C.R., and Saluja, K.K. (1993b). A tutorial on built-in self-test. Part 2: Applications. *IEEE Design & Test of Computers*, 10(2), 69–77.

Albrecht, A.J. (1979). Measuring application development productivity. *Joint SHARE, GUIDE, and IBM Application Development Symposium*, 83–92, October 14–17, Monterey, California, USA.

Anderson, A., Cruess, M., and Wiencek, E. (1989). The binary compatibility standard. *Thirty-fourth IEEE Computer Society International Conference: Intellectual Leverage (COMPCON Spring 89)*, 32–37, February 27, 1989–March 3, 1989.

Backus, J.W., Bauer, F.L., Green, J., Katz, C., McCarthy, J., Perlis, A.J., Rutishauser, H., Samelson, K., Vauquois, B., Wegstein, J.H., van Wijngaarden, A., and Woodger, M. (1960). Report on the algorithmic language ALGOL 60. *Communications of the ACM (CACM)*, 3(5), 299–314.

Backus, J.W., Bauer, F.L., Green, J., Katz, C., McCarthy, J., Naur, P., Perlis, A.J., Rutishauser, H., Samelson, K., Vauquois, B., Wegstein, J.H., van Wijngaarden, A., and Woodger, M. (1963). Revised report on the algorithmic language Algol 60. In *Communications of the ACM (CACM)*, Naur, P. (ed.). 6(1), 1–17.

Bell, J.R. (1973). Threaded code. *Communications of the ACM (CACM)*, 16(6), 370–372.

Carson, J.H. (eds) (1979). Tutorial: Design of microprocessor systems. *Tutorial Week 79*, Institute of Electrical and Electronics Engineering (IEEE), December 10–14, San Diego, California, USA.

Chen, H.-M., Kao, C.-F., and Huang, I.-J. (2002). Analysis of hardware and software approaches to embedded in-circuit emulation of microprocessors. *Seventh Asia-Pacific Computer Systems Architecture Conference (ACSAC'2002)*, Melbourne, Australia. In *Conferences in Research and Practice in Information Technology, 6 (CRPIT'02)*, Lai, F. and Morris, J. (eds), 127–133. *Australian Computer Science Communications*, 24(3).

Cohen, D. (1981). On holy wars and a plea for peace. *IEEE Computer*, 14(10), 48–54. Original: Internet Engineering Note (IEN) 137. USC/ISI (University of Southern California /Information Sciences Institute). April 1, 1980.

Crookes, D. (1983). Teaching assembly-language programming: A high-level approach. *Software & Microsystems*, 2(2), 40–43.

Darche, P. (2000). *Architecture des ordinateurs – Représentation des nombres et codes – Cours avec exercices corrigés*. Éditions Gaëtan Morin.

Darche, P. (2002). *Architecture des ordinateurs – Fonctions booléennes, logiques combinatoire et séquentielle – Cours avec exercices et exemples en VHDL*. Éditions Vuibert.

Darche, P. (2003). *Architecture des ordinateurs – Interfaces et périphériques – Cours avec exercices corrigés*. Éditions Vuibert.

Darche, P. (2004). *Architecture des ordinateurs – Logique booléenne: implémentations et technologies*. Éditions Vuibert.

Darche, P. (2012). *Mémoires à semi-conducteurs: principe de fonctionnement et organisation interne des mémoires vives – Volume 1*. Éditions Vuibert. One of four books selected for the AFISI (Association Française d'Ingénierie des Systèmes d'Information) prize for best computing book 2012.

DaSilva, F., Zorian, Y., Whetsel, L., Arabi, K., and Kapur, R. (2003). Overview of the IEEE P1500 Standard. *2003 International Test Conference (ITC 2003)*, 938–997. Paper 38.1. 30 September–2 October, Charlotte, North Carolina, USA.

Doyle, J. (1985). C – An alternative to assembly programming. *Microprocessors and Microsystems*, 9(3), 124–132.

Ferguson, D.E. (1966). Evolution of the meta-assembly program. *Communications of the ACM (CACM)*, 9(3), 190–196.

Helwig, F. (ed.) (1957). *Coding for the MIT-IBM 704 Computer*. Prepared at the MIT Computation Center by D. Arden, J. McCarthy, S. Best, A. Siegel, F. Corbato, F. Verzuh, F. Helwig, M. Watkins, and M. Weinstein. Massachusetts Institute of Technology, The Technology Press.

Huang, I.-J. and Kao, C.-F. (2000). Exploration of multiple ICEs for embedded microprocessor cores in an SoC chip. *Second IEEE Asia Pacific Conference on ASICs (AP-ASIC 2000)*, 311–314.

Hughes, T.P. and Sawin III, D.H. (1978). Breakpoint design for debugging microprocessor software. *Computer Design*, 17(11), 99–107. Also in Carson 1979, p. 186–194.

Hyde, R. (2010). *The Art of Assembly Language*, 2nd edition. No Starch Press.

IEEE (1985). IEEE Standard for Microprocessor Assembly Language. IEEE Std 694-1985. The Institute of Electrical and Electronics Engineers (IEEE), New York, USA.

IEEE (2000). IEEE Standard VHDL Language Reference Manual. IEEE Std 1076-2000. The Institute of Electrical and Electronics Engineers (IEEE), New York, USA.

IEEE (2003). IEEE Standard for In-system Configuration of Programmable Devices. IEEE Std 1532-2002. Revision of IEEE Std 1532™-2001 (Revision of IEEE Std 1532-2000). The Institute of Electrical and Electronics Engineers (IEEE), New York, USA.

IEEE (2005). 1500™ – IEEE Standard Testability Method for Embedded Core-based Integrated Circuits. IEEE Standard 1500™-2005. The Institute of Electrical and Electronics Engineers (IEEE), New York, USA. Approved June 30, 2005.

IEEE (2013). Standard for Test Access Port and Boundary-Scan Architecture. IEEE Std 1149.1 – 2013 (Revision of IEEE Std 1149.1-2001). The Institute of Electrical and Electronics Engineers (IEEE), New York, USA. Approved February 6, 2013.

IEEE/IEC (2007). IEC 62528 Ed. 1 / IEEE 1500™ Standard Testability Method for Embedded Core-based Integrated Circuits.

IEEE-ISTO (1999). The Nexus 5001 Forum™ Standard for a Global Embedded Processor Debug Interface. IEEE-ISTO Std 5001™-1999.

ISO (1996). Information Technology – Syntactic Metalanguage – Extended BNF. ISO/IEC 14977:1996(E).

Jones, C. (1986). *Programming Productivity: Steps Toward a Science*. McGraw-Hill, New York, USA.

Jones, C. (1998). Sizing up software. *Scientific American*, 279(6), 107–109.

Kao, C.-F., Huang, I.-J., and Chen, H.-M. (2008). Hardware–software approaches to in-circuit emulation for embedded processors. *IEEE Design & Test of Computers*, 25(5), 462–477.

Kernighan, B.W. (1983). A programming language called C. *IEEE Potentials*, 2(4), 26–30.

Kline, B., Maerz, M., and Rosenfeld, P. (1976). The in-circuit approach to the development of microcomputer-based products. *Proceedings of the IEEE*, 64(6), 937–942. Also in Carson 1979, p. 104–109.

Knuth, D.E. (1964). Backus normal form vs. Backus Naur form. *Communications of the ACM (CACM)*, 7(12), 735–736.

Krummel, L. and Schultz, G. (1977). Advances in microcomputer development systems. *IEEE Computer*, 10(2), 13–19. Republished in Carson 1979, p. 87–93.

Maunder, C.M. and Tulloss, R.E. (1990). *The Test Access Port and Boundary-scan Architecture*. IEEE Computer Society Press Tutorial. IEEE Computer Society Press.

McEwan, R. (2002). MPC8xx Using BDM and JTAG. Application Note. AN2387/D. Rev. 0. Motorola Inc., November.

McIlroy, M.D. (1960). Macro instruction extensions of compiler languages. *Communications of the ACM (CACM)*, 3(4), 214–220.

Motorola (1992). DSP56000 Digital Signal Processor Family Manual. Ref. DSP56KFAMUM/AD. Motorola Semiconductor Products Inc.

Navabi, Z. (2010). *Digital System Test and Testable Design: Using HDL Models and Architectures*. Springer-Verlag Inc., New York, USA.

Nicoud, J.-D. (1976). A common assembly language. *Second Euromicro Conference*, 213–220.

Novell (1995a). *System V Interface Definition, Volume 1*, 4th edition. Novell, Inc.

Novell (1995b). *System V Interface Definition, Volume 2*, 4th edition. Novell, Inc.

Novell (1995c). *System V Interface Definition, Volume 3*, 4th edition. Novell, Inc.

Parker, K.P. (2002). *The Boundary-scan Handbook – Analog and Digital*, 2nd edition. Kluwer Academic Publishers.

Poley, S. and Mitchell, G.E. (1956). Symbolic Optimum Assembly Programming (S.O.A.P.). *650 Programming Bulletin*, 1. IBM.

SCO (1996a). System V Application Binary Interface, 4th edition. The Santa Cruz Operation, Inc., California, USA.

SCO (1996b). System V Application Binary Interface. Intel386 Architecture Processor Supplement, 4th edition. The Santa Cruz Operation, Inc., California, USA.

SCO (1997). System V Application Binary Interface, 1st edition. The Santa Cruz Operation, Inc., California, USA.

Smith, J.E. and Nair, R. (2005). *Virtual Machines. Versatile Platforms for Systems and Processes*. Morgan Kaufmann Publishers, Elsevier Inc.

Stiefel, M.L. (1979). Microcomputer development systems. Product Profile. *Mini-Micro Systems*, 12(9), 74. Republished in Carson 1979, p. 78–86.

Stollon, N. (2011). *On-Chip Instrumentation – Design and Debug for Systems on Chip*. Springer Science+Business Media, LLC.

Streib, J.T. (2011). *Guide to Assembly Language: A Concise Introduction*. Springer-Verlag London Ltd.

Vermeulen, B., Stollon, N., Kühnis, R., Swoboda, G., and Rearick, J. (2008). Overview of debug standardization activities. *IEEE Design & Test of Computers*, 25(3), 258–267.

Wang, L.-T., Wu, C.-W., and Wen, X. (eds) (2006). *VLSI Test Principles and Architectures: Design for Testability*. The Morgan Kaufmann Publishers, Elsevier Inc.

Wegner, P. (1968). *Programming Language, Information Structure, and Machine Organization*. McGraw-Hill Inc., New York, USA.

Wegner, P. (1976). Programming languages – The first 25 years. *IEEE Transactions on Computers*, C-25(12), 1207–1225.

Wheeler, D.J. (1950). Programme Organization and Initial Orders for the EDSAC. *Proceedings of the Royal Society of London, Series A, Mathematical and Physical Sciences*, 202(1071), 573–589.

Wilhelm, R. and Maurer, D. (1994). *Les compilateurs – Théorie. construction. génération*. Masson.

Wilkes, M.V. and Renwick, W. (1950). The EDSAC (Electronic Delay Storage Automatic Calculator). *Mathematical Tables and other Aids to Computation (Mathematics of Computation)*, IV(30), 61–65.

Wilkes, M.V., Wheeler, D.J., and Gill, S. (1951). *The Preparation of Programs for an Electronic Digital Computer, with Special Reference to the EDSAC and the Use of a Library of Subroutines*. Addison-Wesley Press, Inc., Cambridge, Massachusetts.

Williams, M.J.Y. and Angell, J.B. (1973). Enhancing testability of large-scale integrated circuits via test points and additional logic. *IEEE Transactions on Computers*, C-22(1), 46–60.

Wirth, N. (1968). PL360, a programming language for the 360 computers. *Journal of the ACM (JACM)*, 15(1), 37–74.

Part 2 (Chapter 3)

Ajanovic, J. and Harriman, D.J. (2000). Method and apparatus for an improved interface between computer components. American patent n° 6145039. Intel Corporation. Filing date: November 3, 1998.

Bradley, D. (2011). A personal history of the IBM PC. *IEEE Computer*, 44(8), 19–25.

Bride, E. (2011). The IBM personal computer: A software-driven market. *IEEE Computer*, 44(8), 34–39.

Ciarcia, S. (1987a). Part 1: AT basics. Build the circuit cellar AT computer. Ciarcia's circuit cellar. *Byte*, 12(10), 115–120.

Ciarcia, S. (1987b). Part 2: Schematic. Build the circuit cellar AT computer. Ciarcia's circuit cellar. *Byte*, 12(11), 135–143.

Croucher, P. (2004). *The BIOS Companion*. Electrocution Technical Publishers.

Dice, P. (2018). *Quick Boot: A Guide for Embedded Firmware Developers*, 2nd edition. De|G Press.

Gayler, W. (1983). *The Apple II® Circuit Description*. Howard W. Sams & Co., Inc.

Goth, G. (2011). IBM PC retrospective: There was enough right to make it work. *IEEE Computer*, 44(8), 26–33, August.

Gwennap, L. (1993). Intel provides PCI chip set for Pentium. New PCI-to-EISA Bridge also works with 486 chip set. *Microprocessor Report (MPR)*, 7(4).

Holden, B., Trodden, J., and Anderson, D. (2008). *HyperTransport 3.1 Interconnect Technology*. MindShare, Inc. MindShare Press.

IBM (1981). Technical Reference. Personal Computer Hardware Reference Library, 1st edition. International Business Machines Corporation.

IBM (1983). Technical Reference. Personal Computer XT Hardware Reference Library, 1st edition (Revised May 1983). International Business Machines Corporation.

IBM (1984a). Technical Reference. Personal Computer Hardware Reference Library, revised edition. International Business Machines Corporation.

IBM (1984b). Technical Reference. Personal Computer Hardware Reference Library, reference 1502494, 1st edition. International Business Machines Corporation.

IBM (1986). Technical Reference. Personal Computer Hardware Reference Library, reference 6183355, 1st edition. International Business Machines Corporation.

IEEE (1994a). IEEE Std 1275.1-1994. IEEE Standard for Boot (Initialization Configuration) Firmware: Instruction Set Architecture (ISA) Supplement for IEEE 1754. IEEE.

IEEE (1994b). IEEE Std 1275.2-1994. IEEE Standard for Boot (Initialization Configuration) Firmware: Bus Supplement for IEEE 1496 (SBus). IEEE.

IEEE (1994c). IEEE Std 1275-1994. IEEE Standard for Boot (Initialization Configuration) Firmware: Core Requirements and Practices. IEEE.

IEEE (1995). IEEE Std 1275.4-1995 (Supplement to IEEE Std 1275-1994). IEEE Standard for Boot (Initialization Configuration) Firmware: IEEE Standard for Boot (Initialization Configuration) Firmware: Bus Supplement for IEEE 896 (Futurebus+®). Approved September 21.

Intel (1993). 82420/82430 PCIset. ISA and EISA Bridges. Databook. Intel Corporation.

Intel (1996). Intel 450KX/GX PCIset. Databook. Intel Corporation.

Intel (2002). Extensible Firmware Interface Specification. Version 1.10. Intel.

Noll, M.J. (1998). System for a Primary BIOS ROM Recovery in a Dual BIOS ROM Computer System. American patent n° 5793943. Application number: 08/688056. Filing date: July 29, 1996.

Opler, A. (1967). Fourth-generation software, the realigment. *Datamation*, 13(1), 22–24.

PCI-SIG (1998). PCI Local Bus Specification, Revision 2.2. PCI-SIG (Special Interest Group).

Singh, G. (2011). The IBM PC: The silicon story. *IEEE Computer*, 44(8), 40–45.

Slater, M. (1992). Intel unveils first PCI chip set. Three-chip set supports high-end 486 ISA systems. *Microprocessor Report (MPR)*, 6(1).

Trodden, J. and Anderson, D. (2003). *HyperTransport™ System Architecture.* MindShare, Inc., Addison-Wesley Developers Press.

UEFI (2017). *Unified Extensible Firmware Interface Specification*, Version 2.7. Unified EFI Forum, Inc.

Index

This index covers all 5 volumes in this series of books.

B

C

D

F

J

L

M

N

O

S

T

U

V

W

Other titles from

iSTE

in

Computer Engineering

2020

DUVAUT Patrick, DALLOZ Xavier, MENGA David, KOEHL François, CHRIQUI Vidal, BRILL Joerg
Internet of Augmented Me, I.AM: Empowering Innovation for a New Sustainable Future

LAFFLY Dominique
TORUS 1 – Toward an Open Resource Using Services: Cloud Computing for Environmental Data
TORUS 2 – Toward an Open Resource Using Services: Cloud Computing for Environmental Data
TORUS 3 – Toward an Open Resource Using Services: Cloud Computing for Environmental Data

LAURENT Anne, LAURENT Dominique, MADERA Cédrine
Data Lakes
(Databases and Big Data Set – Volume 2)

OULHADJ Hamouche, DAACHI Boubaker, MENASRI Riad
Metaheuristics for Robotics
(Optimization Heuristics Set – Volume 2)

SADIQUI Ali
Computer Network Security

VENTRE Daniel
Artificial Intelligence, Cybersecurity and Cyber Defense

2019

BESBES Walid, DHOUIB Diala, WASSAN Niaz, MARREKCHI Emna
Solving Transport Problems: Towards Green Logistics

CLERC Maurice
Iterative Optimizers: Difficulty Measures and Benchmarks

GHLALA Riadh
Analytic SQL in SQL Server 2014/2016

TOUNSI Wiem
Cyber-Vigilance and Digital Trust: Cyber Security in the Era of Cloud Computing and IoT

2018

ANDRO Mathieu
Digital Libraries and Crowdsourcing
(Digital Tools and Uses Set – Volume 5)

ARNALDI Bruno, GUITTON Pascal, MOREAU Guillaume
Virtual Reality and Augmented Reality: Myths and Realities

BERTHIER Thierry, TEBOUL Bruno
From Digital Traces to Algorithmic Projections

CARDON Alain
Beyond Artificial Intelligence: From Human Consciousness to Artificial Consciousness

HOMAYOUNI S. Mahdi, FONTES Dalila B.M.M.
Metaheuristics for Maritime Operations
(Optimization Heuristics Set – Volume 1)

JEANSOULIN Robert
JavaScript and Open Data

PIVERT Olivier
NoSQL Data Models: Trends and Challenges
(Databases and Big Data Set – Volume 1)

SEDKAOUI Soraya
Data Analytics and Big Data

SALEH Imad, AMMI Mehdi, SZONIECKY Samuel
Challenges of the Internet of Things: Technology, Use, Ethics
(Digital Tools and Uses Set – Volume 7)

SZONIECKY Samuel
Ecosystems Knowledge: Modeling and Analysis Method for Information and Communication
(Digital Tools and Uses Set – Volume 6)

2017

BENMAMMAR Badr
Concurrent, Real-Time and Distributed Programming in Java

HÉLIODORE Frédéric, NAKIB Amir, ISMAIL Boussaad, OUCHRAA Salma, SCHMITT Laurent
Metaheuristics for Intelligent Electrical Networks
(Metaheuristics Set – Volume 10)

MA Haiping, SIMON Dan
Evolutionary Computation with Biogeography-based Optimization
(Metaheuristics Set – Volume 8)

PÉTROWSKI Alain, BEN-HAMIDA Sana
Evolutionary Algorithms
(Metaheuristics Set – Volume 9)

PAI G A Vijayalakshmi
Metaheuristics for Portfolio Optimization
(Metaheuristics Set – Volume 11)

2016

BLUM Christian, FESTA Paola
Metaheuristics for String Problems in Bio-informatics
(Metaheuristics Set – Volume 6)

DEROUSSI Laurent
Metaheuristics for Logistics
(Metaheuristics Set – Volume 4)

DHAENENS Clarisse and JOURDAN Laetitia
Metaheuristics for Big Data
(Metaheuristics Set – Volume 5)

LABADIE Nacima, PRINS Christian, PRODHON Caroline
Metaheuristics for Vehicle Routing Problems
(Metaheuristics Set – Volume 3)

LEROY Laure
Eyestrain Reduction in Stereoscopy

LUTTON Evelyne, PERROT Nathalie, TONDA Albert
Evolutionary Algorithms for Food Science and Technology
(Metaheuristics Set – Volume 7)

MAGOULÈS Frédéric, ZHAO Hai-Xiang
Data Mining and Machine Learning in Building Energy Analysis

RIGO Michel
Advanced Graph Theory and Combinatorics

2015

BARBIER Franck, RECOUSSINE Jean-Luc
COBOL Software Modernization: From Principles to Implementation with the BLU AGE® Method

CHEN Ken
Performance Evaluation by Simulation and Analysis with Applications to Computer Networks

CLERC Maurice
Guided Randomness in Optimization
(Metaheuristics Set – Volume 1)

DURAND Nicolas, GIANAZZA David, GOTTELAND Jean-Baptiste, ALLIOT Jean-Marc
Metaheuristics for Air Traffic Management
(Metaheuristics Set – Volume 2)

MAGOULÈS Frédéric, ROUX François-Xavier, HOUZEAUX Guillaume
Parallel Scientific Computing

MUNEESAWANG Paisarn, YAMMEN Suchart
Visual Inspection Technology in the Hard Disk Drive Industry

2014

BOULANGER Jean-Louis
Formal Methods Applied to Industrial Complex Systems

BOULANGER Jean-Louis
Formal Methods Applied to Complex Systems:Implementation of the B Method

GARDI Frédéric, BENOIST Thierry, DARLAY Julien, ESTELLON Bertrand, MEGEL Romain
Mathematical Programming Solver based on Local Search

KRICHEN Saoussen, CHAOUACHI Jouhaina
Graph-related Optimization and Decision Support Systems

LARRIEU Nicolas, VARET Antoine
Rapid Prototyping of Software for Avionics Systems: Model-oriented Approaches for Complex Systems Certification

OUSSALAH Mourad Chabane
Software Architecture 1
Software Architecture 2

PASCHOS Vangelis Th
Combinatorial Optimization – 3-volume series, 2nd Edition
Concepts of Combinatorial Optimization – Volume 1, 2nd Edition
Problems and New Approaches – Volume 2, 2nd Edition
Applications of Combinatorial Optimization – Volume 3, 2nd Edition

QUESNEL Flavien
Scheduling of Large-scale Virtualized Infrastructures: Toward Cooperative Management

RIGO Michel
Formal Languages, Automata and Numeration Systems 1: Introduction to Combinatorics on Words
Formal Languages, Automata and Numeration Systems 2: Applications to Recognizability and Decidability

SAINT-DIZIER Patrick
Musical Rhetoric: Foundations and Annotation Schemes

TOUATI Sid, DE DINECHIN Benoit
Advanced Backend Optimization

2013

ANDRÉ Etienne, SOULAT Romain
The Inverse Method: Parametric Verification of Real-time Embedded Systems

BOULANGER Jean-Louis
Safety Management for Software-based Equipment

DELAHAYE Daniel, PUECHMOREL Stéphane
Modeling and Optimization of Air Traffic

FRANCOPOULO Gil
LMF — Lexical Markup Framework

GHÉDIRA Khaled
Constraint Satisfaction Problems

ROCHANGE Christine, UHRIG Sascha, SAINRAT Pascal
Time-Predictable Architectures

WAHBI Mohamed
Algorithms and Ordering Heuristics for Distributed Constraint Satisfaction Problems

ZELM Martin *et al.*
Enterprise Interoperability

2012

ARBOLEDA Hugo, ROYER Jean-Claude
Model-Driven and Software Product Line Engineering

BLANCHET Gérard, DUPOUY Bertrand
Computer Architecture

BOULANGER Jean-Louis
Industrial Use of Formal Methods: Formal Verification

BOULANGER Jean-Louis
Formal Method: Industrial Use from Model to the Code

CALVARY Gaëlle, DELOT Thierry, SÈDES Florence, TIGLI Jean-Yves
Computer Science and Ambient Intelligence

MAHOUT Vincent
Assembly Language Programming: ARM Cortex-M3 2.0: Organization, Innovation and Territory

MARLET Renaud
Program Specialization

SOTO Maria, SEVAUX Marc, ROSSI André, LAURENT Johann
Memory Allocation Problems in Embedded Systems: Optimization Methods

2011

BICHOT Charles-Edmond, SIARRY Patrick
Graph Partitioning

BOULANGER Jean-Louis
Static Analysis of Software: The Abstract Interpretation

CAFERRA Ricardo
Logic for Computer Science and Artificial Intelligence

HOMES Bernard
Fundamentals of Software Testing

KORDON Fabrice, HADDAD Serge, PAUTET Laurent, PETRUCCI Laure
Distributed Systems: Design and Algorithms

KORDON Fabrice, HADDAD Serge, PAUTET Laurent, PETRUCCI Laure
Models and Analysis in Distributed Systems

LORCA Xavier
Tree-based Graph Partitioning Constraint

TRUCHET Charlotte, ASSAYAG Gerard
Constraint Programming in Music

VICAT-BLANC PRIMET Pascale *et al.*
Computing Networks: From Cluster to Cloud Computing

2010

AUDIBERT Pierre
Mathematics for Informatics and Computer Science

BABAU Jean-Philippe *et al.*
Model Driven Engineering for Distributed Real-Time Embedded Systems

BOULANGER Jean-Louis
Safety of Computer Architectures

MONMARCHE Nicolas *et al.*
Artificial Ants

PANETTO Hervé, BOUDJLIDA Nacer
Interoperability for Enterprise Software and Applications 2010

SIGAUD Olivier *et al.*
Markov Decision Processes in Artificial Intelligence

SOLNON Christine
Ant Colony Optimization and Constraint Programming

AUBRUN Christophe, SIMON Daniel, SONG Ye-Qiong *et al.*
Co-design Approaches for Dependable Networked Control Systems

2009

FOURNIER Jean-Claude
Graph Theory and Applications

GUEDON Jeanpierre
The Mojette Transform / Theory and Applications

JARD Claude, ROUX Olivier
Communicating Embedded Systems / Software and Design

LECOUTRE Christophe
Constraint Networks / Targeting Simplicity for Techniques and Algorithms

2008

BANÂTRE Michel, MARRÓN Pedro José, OLLERO Hannibal, WOLITZ Adam
Cooperating Embedded Systems and Wireless Sensor Networks

MERZ Stephan, NAVET Nicolas
Modeling and Verification of Real-time Systems

PASCHOS Vangelis Th
Combinatorial Optimization and Theoretical Computer Science: Interfaces and Perspectives

WALDNER Jean-Baptiste
Nanocomputers and Swarm Intelligence

2007

BENHAMOU Frédéric, JUSSIEN Narendra, O'SULLIVAN Barry
Trends in Constraint Programming

JUSSIEN Narendra
A TO Z OF SUDOKU

2006

BABAU Jean-Philippe *et al.*
From MDD Concepts to Experiments and Illustrations – DRES 2006

HABRIAS Henri, FRAPPIER Marc
Software Specification Methods

MURAT Cecile, PASCHOS Vangelis Th
Probabilistic Combinatorial Optimization on Graphs

PANETTO Hervé, BOUDJLIDA Nacer
Interoperability for Enterprise Software and Applications 2006 / IFAC-IFIP I-ESA'2006

2005

GÉRARD Sébastien *et al.*
Model Driven Engineering for Distributed Real Time Embedded Systems

PANETTO Hervé
Interoperability of Enterprise Software and Applications 2005

Printed in the USA/Agawam, MA
September 16, 2026

910661.019